Environmental Footprints and Eco-design of Products and Processes

Series Editor

Subramanian Senthilkannan Muthu, Green Story Inc.
Toronto, Canada

Indexed by Scopus

This series aims to broadly cover all the aspects related to environmental assessment of products, development of environmental and ecological indicators and eco-design of various products and processes. Below are the areas fall under the aims and scope of this series, but not limited to: Environmental Life Cycle Assessment; Social Life Cycle Assessment; Organizational and Product Carbon Footprints; Ecological, Energy and Water Footprints; Life cycle costing; Environmental and sustainable indicators; Environmental impact assessment methods and tools; Eco-design (sustainable design) aspects and tools; Biodegradation studies; Recycling; Solid waste management; Environmental and social audits; Green Purchasing and tools; Product environmental footprints; Environmental management standards and regulations; Eco-labels; Green Claims and green washing; Assessment of sustainability aspects.

Subramanian Senthilkannan Muthu
Editor

Circular Business Models in the Apparel Industry

Springer

Editor
Subramanian Senthilkannan Muthu
Chief Sustainability Officer
Green Story, Canada
Tung Chung, Hong Kong

ISSN 2345-7651 ISSN 2345-766X (electronic)
Environmental Footprints and Eco-design of Products and Processes
ISBN 978-3-031-80192-1 ISBN 978-3-031-80193-8 (eBook)
https://doi.org/10.1007/978-3-031-80193-8

This Springer imprint is published by the registered company Springer Nature Switzerland AG
The registered company address is: Gewerbestrasse 11, 6330 Cham, Switzerland

This book is dedicated to:
The lotus feet of my beloved Lord
Pazhaniandavar
My beloved late Father
My beloved Mother
My beloved Wife Karpagam and Daughters—
Anu and Karthika
My beloved Brother
Last but not least
To everyone working in various industrial
spaces to make the world "SUSTAINABLE"

Contents

Introduction to the Book- Circular Business Models in the Apparel Industry

Subramanian Senthilkannan Muthu

Circular business models in the apparel industry represent a paradigm shift from the traditional linear "take-make-dispose" approach to a more sustainable and regenerative system. These models aim to extend the life cycle of clothing, maximize resource efficiency, and minimize waste by keeping products and materials in use for as long as possible. As the fashion industry grapples with its significant environmental impact, circular business models offer a promising solution to reduce resource consumption, cut carbon emissions, and create new revenue streams. The adoption of circular practices in fashion encompasses various strategies, including resale, rental, repair, and remaking. These models not only address environmental concerns but also cater to changing consumer preferences for more sustainable and affordable fashion options. With the potential to claim 23 per cent of the global fashion market by 2030 and generate USD 700 billion in value, circular business models are poised to transform the industry. However, successful implementation requires a holistic approach that aligns innovation, policy, and consumer behaviour to create a truly sustainable fashion ecosystem. This chapter will present an overview of the entire book.

The second chapter deals with the circular economy solutions for sustainability in textile and silk sector. The textile sector has a considerable environmental impact throughout its value chain. It has arisen as an important player in global manufacturing, producing a substantial 113 million tonnes of fibre in 2021. This figure is anticipated to rise to 149 million tonnes by the year 2030. Consequently, recycled textiles currently contribute less than 1 per cent to the global fibre market, underscoring the urgent necessity for business models for circularity within the industry of textile. Circular economy (CE) is acknowledged as the most effective means to attain sustainability and circularity. Within this framework, the industry is a central focus of

S. S. Muthu (✉)
Chief Sustainability Officer, Green Story, Canada, Tung Chung, Hong Kong

S. S. Muthu (ed.), *Circular Business Models in the Apparel Industry*, Environmental Footprints and Eco-design of Products and Processes,
https://doi.org/10.1007/978-3-031-80193-8_1

the forthcoming strategy related to Green Deal, regarding making Europe climate-neutral by 2050. This focus is echoed in the European Union (EU) textiles strategy, which aims to strengthen the market for circular and sustainable textiles within the EU. While numerous studies have explored the establishment of business models related to circularity, there is presently a need to shift attention to the silk industry, given its significance in the textile sector and its potential contribution to circularity. Educating consumers, who play an important part in attaining circularity in the silk industry, is essential. This chapter aims to gather insights from various perspectives, including the primary and secondary activities performed in the silk value chain and the implementation of circular strategies for the management of textile waste, and the strategies involve various methods comprising reusing, repairing, renting, reselling, refurbishing, and upcycling. These strategies can help attain circularity and lessen waste and the industry's negative environmental impact. Furthermore, the revolutionary methods for silk waste recycling revel how the industries can convert silk waste into the valuable products like new clothing and reinforcement composite and how this will minimize the impact of textile waste and boost the economic value.

The third chapter revolves around the taxonomy of sustainable and circular apparel business models. The apparel industry is characterized by an increasing overconsumption and a prevalent environmental linear production system that employs non-renewable resources and intensive water use. To address this challenge, the concept of circular economy (CE) has recently gained considerable attention in order to facilitate the transition towards more sustainable fashion systems. New circular business models have been launched in the fashion industry and often require a paradigm shift in how fashion garments are produced, consumed, and disposed. This study describes the main circular business models related to the consumption field that are spreading in the apparel market. The aim is to present an academic taxonomy of circular consumption business models in the apparel sector, illustrating both the most widespread ones and successful business cases.

The fourth chapter deals with the circular economy dynamics in the apparel industry, with respect to the emergence of "Retrovation". The apparel industry is witnessing a significant transformation towards circularity in response to environmental concerns, with an original trend being the resurgence of traditional practices and processes under innovative contexts, termed "retrovation". This chapter explores the intricate interplay between circularity and retrovation, aiming to illustrate how a retrovation-driven circular approach can foster sustainable transformations. It examines the convergence of craftsmanship, traditional production methods, and innovation with contemporary sustainability. The empirical analysis is based on a case study that investigates the application of retrovative activities for sustainability within the apparel industry. Specifically, it illustrates the revitalization of traditional practices by Rifò, a successful business within the Prato textile industrial district in Italy, spotlighting the original and innovative resurgence of *cenciaioli*'s artisanal abilities. *Cenciaioli* are renowned for their expertise in selecting and repurposing textile scraps and waste, and they epitomize the ethos of retrovation. Originating from a tradition of pragmatism, they now lead sustainability innovation by integrating traditional knowledge with current attention to environmental sustainability. By examining the fusion of circularity and retrovation within the apparel industry, this study illuminates the potential of retrovation in connecting cultural heritage practices and innovation for increasing the sustainability of the apparel industry towards the implementation of an effective circular apparel ecosystem.

The fifth chapter deals with modelling social circular economy business models (CEBMs) in the fashion industry with the Atelier Riforma case study. The crucial role of circular economy business models (CEBMs) in mitigating environmental degradation and decoupling economic growth from material use has been globally recognized in the literature across every economic sector. However, while the environmental benefits of circular economy (CE) are well documented, its social value has only recently begun to receive scholarly attention. Indeed, how organizations with a circular business model and an integrated social mission minimize negative environmental impacts while amplifying positive social outcomes remains unclear. This chapter aims to theorize the emergent social circular economy (SCE) by analysing a case study focused on Atelier Riforma, an Italian innovative start-up with a social vocation in the apparel sector, and modelling novel integrated CEBMs. Employing semi-structured interviews, direct observations, and internal documents, this study sheds light on the operationalization of CE principles intertwined with a social mission. It discusses, in terms of material, financial, and impact flows, how SCE strategies in a sector currently highly unsustainable as the fashion industry may create (financial) value by recovering garments and simultaneously creating both social (i.e. labour value) and environmental (e.g. the reduction of material) values. This chapter, hence, enriches the discourse on CE and the social and solidarity economy (SSE) by illustrating how innovative platform business models may address and assess the dual objectives of environmental sustainability and social equity.

The sixth chapter deals with the rise of circular fashion for sustainable apparel retailing. Circular fashion aims to change the traditional fashion model into a sustainable system that reduces waste, reuses garments, and regenerates natural systems. Different business models exist like closed-loop recycling, product-as-a-service (PaaS), design for longevity, and take-back-give-back schemes for elevating circular fashion. These models are sustainable systems that reduce waste, promote resource efficiency, and extend product use. The growth is witnessed in "circular fashion" due to increased demand for sustainability, government initiatives to promote sustainable products, economic opportunities, etc. There are also various new business models other than the available and existing ones which reconceptualize ownership and lead people to follow "circular fashion". However, the rise in circular fashion does have challenges and this chapter discusses those challenges along with the opportunities. The chapter also discusses how consumer behaviour impacts the circular fashion concept and thereby sustainability. The various ways of increasing consumer awareness towards sustainability are also mentioned in this chapter. The policy and regulatory frameworks play important roles in promoting circular fashion as they provide the roadmap for the future of the country and the world. So their impact is also detailed in the chapter. The chapter also delves into the economic viability of circular fashion. The whole process involves building a sustainable and transparent supply chain, and the ways to build them are discussed along with the barriers in "circular fashion". This chapter gives a detailed view of how the rise of circular fashion leads to sustainable apparel retailing.

The seventh chapter deals with materials, design, and technology in the circular fashion for sustainability. Circular fashion is a sustainability initiative taken by the fashion industry to reduce the ill-effects of fashion on the environment. The process of circular fashion involves utilizing the produced material as many times as possible in different forms before throwing them as waste in landfills. It is imperative to employ the circular fashion principles to safeguard the environment and also protect the environment for a better future. This chapter explores the concept of circular fashion as a sustainable alternative to the traditional linear fashion model. By tracing the evolution of the fashion industry and its environmental impact, the research highlights the urgent need for a circular approach that prioritizes waste reduction, resource efficiency, and product longevity. This chapter delves into the core principles of circular fashion, encompassing reduction, reuse, recycle, and redesign strategies. Furthermore, it explores a wide range of sustainable materials, innovative recycling technologies, and emerging bio-fabrication methods. The study also emphasizes the importance of designing durable and long-lasting apparel through careful fabric selection, construction techniques, and timeless design aesthetics. The fashion industry's sustainability evolution and impact on the environment are also discussed in this chapter. By combining these elements, the chapter provides details on the growth of circular fashion and provides valuable information for industry stakeholders, policymakers, and consumers to foster a more sustainable fashion future.

I take this opportunity to thank all the contributors for their earnest efforts to bring out this book successfully. I am sure readers of this book will find it very useful.

With best wishes,
Dr. Subramanian Senthilkannan Muthu.

Circular Economy Solutions for Sustainability in Textile and Silk Sector

Rabia Hassan, Federica Acerbi, Sergio Terzi, and Paolo Rosa

1 Introduction

The movement toward a circular economy in the textiles field is largely a reaction to the dissemination of reports revealing the detrimental effects of fashion consumption (Ellen MacArthur Foundation, 2017). The industry of textile arises as an important player in global manufacturing, producing a substantial 113 million tons of fiber in 2021. This figure is anticipated to rise to 149 million tons by the year 2030. Consequently, the global recycled fiber market for textiles, including pre-consumer and post-consumer materials, accounted for only 1% (Textile Exchange, 2022b). With regard to usage, compared to 15 years ago, consumers now buy double the quantity of clothes but typically retain them for only half the time. This increased production and consumption, coupled with shorter usage periods, results in a significant volume of textile waste after consumer use (Cao et al., 2022). During past few decenniums, there has been a reported 40% growth in the amount of clothing purchased in the European Union (EU). Approximately, Europeans now consume 26 kg of textiles per person annually and dispose 11.3 kg per person each year; given the concerning state of the climate, this trend is alarming (Riemens et al., 2021). Annually, in the EU, approximately 2 million tons of textiles are discarded. Consequently, roughly one truck of textile garbage is either incinerated or buried in landfills per second (Papamichael et al., 2023).

Various governments worldwide are implementing economic policies designed to incentivize European consumers and businesses to transition toward a more

R. Hassan (✉) · F. Acerbi · S. Terzi · P. Rosa
Department of Management, Economics, and Industrial Engineering, Politecnico di Milano, Milan, Italy
e-mail: rabia.hassan@polimi.it; federica.acerbi@polimi.it; sergio.terzi@polimi.it; paolo1.rosa@polimi.it

S. S. Muthu (ed.), *Circular Business Models in the Apparel Industry*, Environmental Footprints and Eco-design of Products and Processes,
https://doi.org/10.1007/978-3-031-80193-8_2

comprehensive and resilient circular economy (CE), prioritizing sustainable resource management. It targets to eradicate waste resulting from the materials production, equipment and procedures to assist the formation business models that are sustainable (Kirchherr et al., 2017). In pursuit of a more competitive and sustainable economy, the European Union (EU) is introducing a variety of regulatory measures to reduce resource use and production waste while promoting health and productivity. The CE concept aims to change the current business model, based on ecology, environment, and business economics studies. The system facilitates the recovering, reusing, and recycling of services and products, extending their valuable lifespan and extending the chain of production. Furthermore, the concept seeks to enhance both economic profitability and societal well-being by ensuring the longevity of technological and biological resources during the value chain of goods consumed by the population to prioritize resource optimization (Esbeih et al., 2021).

However, circular economy is based on three core principles, with the first emphasizing the importance of preserving and enhancing the environment. The second principle involves optimizing resource performance by maximizing material usage, designing for remanufacturing and longevity, and recycling to maintain materials in circulation. Third, it involves removing negative outwardness from the design process through the implementation of efficient systems (Ellen MacArthur Foundation, 2013a, b). In this way, textile and clothing industries can attain circularity throughout the whole lifespan of products. This chapter begins with an introduction that underscores the significance of the circular economy within the textile industry. It then examines textile waste and its environmental impact, comparing various natural fibers. The chapter delves into the silk industry, focusing on market dynamics, silk waste (SW), and the silk value chain, including both primary and supportive activities, to emphasize the importance of silk fiber production. It also discusses circular strategies, placing particular emphasis on methods and techniques for recycling silk to achieve circularity in the industry. Additionally, the chapter explores the application of silk waste across different sectors, highlighting its environmentally friendly benefits. The conclusion summarizes the key findings and proposes potential directions for future research.

1.1 Circular Economy

The definition of a CE, as established by the Ellen MacArthur Foundation (2013a), emphasizes an industrialized structure intentionally designed to be restorative. From certain viewpoints, the essence of a CE involves creating a material's seamless and closed cycle, along with optimizing the use of energy and raw materials throughout different stages. The key dissimilarity among a circular and linear economy is how long materials and resources maintain their value. In a CE, the focus is on keeping these materials in use as functional products for the maximum possible

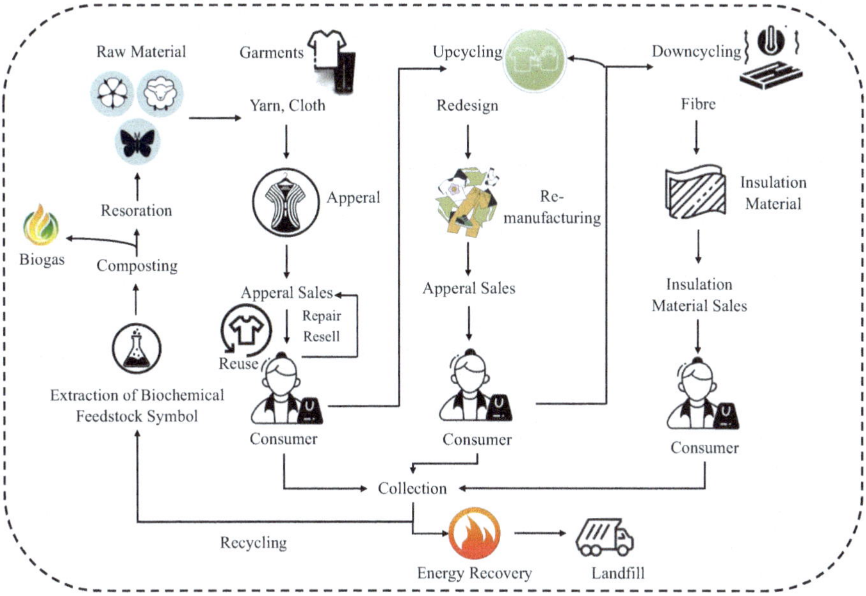

Fig. 1 Textile circulation model (Author's Creation). (Source: Ellen MacArthur Foundation, 2013a)

time. The value chain in a linear economy, starts from the production or extraction of materials like fiber and extends to the product being sold to consumers. Once purchased, the product is used until it no longer functions or is no longer wanted by the owner for various reasons. When this occurs, the product is replaced and the old one is discarded, ending its life cycle (Fernandes et al., 2023).

In the CE value chain of textile, the entire life cycle of a product includes both forward and reverse elements as illustrated in Fig. 1. In this model, the reverse value chain plays a critical role. It adds value by effectively reintegrating previously used products and materials back into the economic system. This integration occurs through a sequential process that includes the collection of used items, their maintenance, and various forms of reprocessing such as reusing, recovering, recycling, and remanufacturing. Each phase is designed to utility maximization and waste minimization of resources, reflecting the core principles of a circular economy (Dragomir & Dumitru, 2022).

The objective of CE is to prolong a product's lifespan by emphasizing durable design, repairability, and enabling transforming or sharing among users when the real owner no longer needs it. The aim is to extend the period of active usage for each product. When a product reaches its end life, materials can be converted into useful resources, which are valuable via recycling; these recycled materials can then be used in manufacturing new or other products. This method insures efficient repurposing and extraction of materials, maintaining their value within the CE.

2 Textile Waste

Textile waste consists of raw materials, fabrics, and clothing or finished items that are no longer deemed useful or functional. Textile waste comprises the generated waste across the whole textile supply chain, encompassing activities from designing and sourcing to production, distribution, and retail. Additionally, it encompasses consumer waste resulting from the use and disposal process, involving the purchasing, wearing, damaging, and eventual discarding of fashion and textile products by consumers. Textile waste specifically arises during three stages. First, textile waste of production (post-industrial waste) consists of fibers, yarn, flocks, cloth scraps, sweepings, cut-offs fabric, roll ends of fabric, and selvage, all of which are primarily generated during the textile and clothing manufacturing processes in factories (Kliś, 2023). Second, pre-consumer waste arises from the retailing of clothing and textiles in both offline and online stores. This includes items that are damaged, defective, or remain unsold (Haq & Alam, 2023). Third, post-consumer wastes are those that consumers have discarded after use. This category includes items that are worn-out, no longer fit, damaged, or no longer align with current trends (Bahl et al., 2023).

2.1 Post-industrial and Pre-consumer Waste

In Europe, waste related to post-industrial amounted to 1.3 Mt. per year, accounting for 37% of the total waste. Within this, 8% was attributed to yarn production, 13% to fabric production, and 20% to finished textile production (Huygens et al., 2023). The anticipated losses of manufacturing are consistent with the evidence presented in the textile strategy of EU, which indicates that 25–40% of fabric in factories either remains unused or is discarded as wastage (Aus et al., 2021). An evaluation of post-industrial waste in Lithuania, revealed by Dobilaite et al. (2017), indicates that cutting waste constitutes around 20–25% of the entire material's amount utilized for fabrication. An analysis of the Swiss textile industry found that 9.5% of the total textile production consisted of yarn and fabric residues (Schmutz & Som, 2022).

Fabric production requires the use of yarns wound around cones, and to ensure proper functioning of the fabric-producing machine, a specific tension level is required. However, maintaining this tension becomes challenging as the yarn approaches its end. Consequently, a minor portion of yarn is left on the cones, ultimately becoming waste. Another form of inherent waste in manufacturing textile is edge trims, which are generated throughout the fabric weaving and afterward during finishing processes of fabric. Sometimes, errors can occur throughout the manufacturing process of yarn or fabric. This may arise from factors like substandard raw materials, improper initial winding of yarn on the cones, or incorrect machine configurations. Additionally, it was noted that there could be slight color variations among two distinctive batches of fabric, which might not meet the designer's expectations. In such cases, complete fabric rolls may end up being wasted (Schmutz & Som, 2022).

However, the pre-consumer waste amounted to 0.33 Mt. per year (Huygens et al., 2023). Pre-consumer waste, which includes products that have not reached consumers, was estimated to constitute around 3–5% in 2020 (McKinsey & Company, 2022). Among unsold products, 70% were freighted outside the EU, while the residual 30% were reintroduced into the EU retail market (Huygens et al., 2023).

2.2 *Post-consumer Waste*

The post-consumer textiles collection separately denotes the predominant fraction of textile waste, amounting to an overall of 10.9 Mt. per year from which it is estimated that around 8 Mt. of post-consumer waste per year goes to incineration or landfill (Huygens et al., 2023). Around 40% of post-consumer waste is sent to second-hand markets in countries outside of Europe, while the remainder undergoes sorting to identify high-demand garments for resale within Europe or other developed nations (McKinsey & Company, 2022). In the analysis conducted by Malinverno et al., (2023) on post-consumer waste from Swiss workwear, it was found that 41% of the waste is reused, indicating a significant portion of the materials being repurposed. Additionally, 40% of the waste is incinerated, likely for energy recovery, while 19% undergoes open-loop recycling, suggesting efforts to extract the value of the materials through recycling processes. Similarly, in China, 40% of post-consumer textile waste lacking utility value requires refibering, while about 15% of clothing undergoes disinfection treatment before being exported to regions like Africa (Xu et al., 2022).

3 Textile Environmental Impact

Raw material extraction involves growing and harvesting natural fibers. The textile manufacturing process, encompassing weaving, knitting, dyeing, finishing, and bleaching, demands a substantial amount of energy. At this stage, fibers are naturally converted into yarn and subsequently made into textile fabrics. However, at every stage of the natural fiber manufacturing, there is an impact on the environment; for example, the raw material extraction stage requires enormous quantities of water, land, energy, fertilizers, and pesticides. The procedure of spinning has more negative impact on the environment, which includes climate change potential, acidification, and abiotic depletion. Similarly, the process of knitting and weaving consumes high electricity, whereas, dyeing and finishing stages of products also have an adverse environmental effect. In view of all these factors, the environmental comparison of cotton, jute, and silk natural fiber are conducted in the below subsections. The primary reasons for selecting these three raw materials are their prevalence and market dominance. Cotton is a predominant fabric in the textile sector, with most of the clothing being made from it. Similarly, jute holds a significant position in the market (Todor et al., 2019). Although, silk has fewer shares in the market, it is still an important fiber with approximately 300,000 households' production (Moazzem et al., 2021).

3.1 Cotton

When organic cotton gained traction as an environmentally friendly substitute for traditional cotton, there was a 31% rise in demand for organic cotton products in 2022 as compared to the 2021 (Textile Exchange, 2022a). Organic cotton is cultivated without the application of chemical substances like fertilizers, insecticides, weed killers, and developmental regulators (Ingram, 2002). In a study by Fidan et al. (2021), a comparison using life cycle assessment (LCA) indicated that fabric denim derived from organic cotton demonstrated lower impact on environment scores across all categories compared to denim sourced from conventional cotton. Şener Fidan et al. (2023) detected a reduction in the whole impact of global warming potential (GWP), shifting from 20.11 kg CO_2 eq. to 19.48 kg CO_2 eq., when shifting from conventional to organic cotton. Moreover, the adoption of organic cotton resulted in a 47.6% rise in the impact category of freshwater aquatic ecotoxicity potential (FAEP).

3.2 Jute

Jute and fibers like hemp have been attracting increasing attention for applications in industry (Ahmed et al., 2022), due to their enhanced properties related to mechanical, relative to other fibers which are natural (Todor et al., 2019). To effectively use plant fibers in different applications, it is essential to recognize how they interrelate with other materials (Marcuello et al., 2023). Furthermore, jute is better for the environment than cotton because it needs a lot lesser amount of water to cultivate and has lesser influence on human health and acidification. When it comes to eutrophication, acidification, water depletion, and the accumulation of energy demands, jute is less harmful to the environment when compared to cotton (La Rosa & Grammatikos, 2019).

3.3 Silk

Silk is recognized for its elegant appearance and biodegradability, but silk-based products influence the environment (Barcelos et al., 2020). Munasinghe et al. (2021) underlined that the production of silk in India has the largest effect toward atmosphere, with the energy consumption of 1467.3 MJ/kg, carbon monoxide emissions of 52.5 kg CO_2 equivalent/kg, and similarly water consumption of 26,700 L/kg. Previous studies have shown that the cultivation of mulberries causes a serious threat to the environment when it comes to freshwater, soil ecology, and human toxicity in the silk industry. The primary role of these involvements is associated with the management of soil, which contains the use of organic fertilizers or phosphorus, which tends to emit metal emissions into the soil and water. The influence of mulberry cultivation on water resources is another reason for consideration (Barcelos et al., 2020). Despite its reputation for elegance and durability, further investigation is required to comprehend fully the environmental effects of silk.

Figure 2 illustrates the environmental effects comparison of jute, cotton and silk in the agricultural, spinning, weaving, and yarn processes (Gonzalez et al., 2023). Compared to many natural fibers, silk stands out the most among all categories. It has a negative impact on the ecotoxicity of soil with a significant impact on fabric production, agriculture, yarn, and spinning and weaving processes, resulting in 26.47910 kg of 1,4-DCB, 85.33574 kg of 1,4-DCB, and 94.59827 kg of 1,4-DCB, respectively. Silk production reveals a significant impact on the global warming category during agricultural, yarn, and spinning and weaving process, marked with a value of 18.66263 kg CO_2 eq, 58.73085 kg CO_2 eq, and 65.92711 kg CO_2 eq, respectively. These figures are drastically greater than the impact of global warming among all other fibers studied (Gonzalez et al., 2023).

Analysis of different natural materials indicates the importance of the environment regarding silk production. This demonstrates the importance of research and circulation efforts, especially in the silk industry. Focusing on silk materials furthers

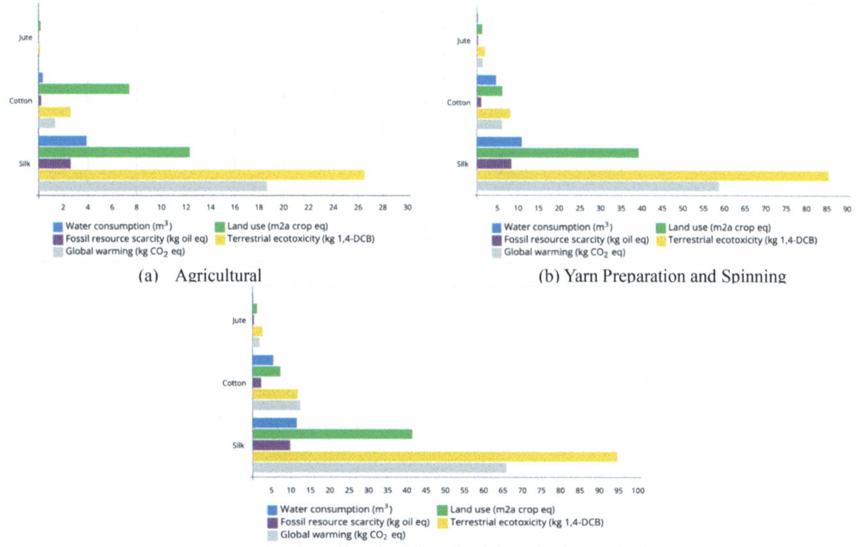

(c) Weaving

Fig. 2 Environmental impacts comparison (Author's Creation). (**a**) Agricultural. (**b**) Yarn preparation and spinning. (**c**) Weaving. (Source: Gonzalez et al., 2023)

the goal of a sustainable textile industry by allowing us to better understand and solve the environmental problems they create. Below, a detailed analysis of the silk industry and value chain is reported.

4 Silk Industry

Bombyx mori occupies a special position among insects and has been widely studied in scientific research as the best representative of the Lepidopteran species. Its important role in various investigations underscores its significance in advancing our wisdom of insect biology and related fields (Panthee et al., 2017). Silkworms are economically important due to their role in the production of silk, which has been a valuable product for centuries. In addition, silk farming is an income source for countless individuals around the world and contributes to the economies of many regions. It provides significant benefits to people due to various uses in practical life. These applications range from its utilization as a model organism for medical research to its various other practical applications (Cappellozza et al., 2022). Sericulture is important for fiber production, which adds value to many industries with its many uses. These include the production of clothes, bags, shoes, wallpaper, and other fabrics (Gjurašić & Đurović, 2022).

The reputation of silk as a symbol of luxury and elegance is rooted in its exceptional qualities and versatility make it the "Queen of fibers". Silk stands out among natural materials for its exceptional durability, resistance to tearing, and high heat resistance, while also possessing a delicate and soft texture. The unique beauty and delicateness of silk fabrics are unmatched by any other material, offering breathability that keeps the skin cool during summer and warm during winter (Babu, 2018). The remarkable properties of silk contribute to its enduring value, which has been acknowledged both in the past and continues to be recognized today.

Silk production constitutes approximately 0.24% of total fiber utilization, amounting to roughly 202,000 metric tons each year. However, roughly 99% of silk production is obtained from *Bombyx mori* (Andadari et al., 2022). Accurately determining the global value of silk products is difficult due to the scarcity of reliable data, as most countries depend on imports of either raw materials or finished silk goods. China is the main silk raw material producer, contributing to 50% of the world's production. India follows with 15%, while Uzbekistan and Brazil produce 3% and 2.5%, respectively. The worldwide raw silk's annual volume averages around 80,000 tons, with China accounting for about 70% of this total. Supported by government initiatives, the silk industry is rapidly growing in Azerbaijan, Kazakhstan, Uzbekistan, and Tajikistan. Although Europe has produced some raw silk, the demand is significant, requiring import of materials outside the EU (Sashina & Yakovleva, 2023).

Despite its advantages as a superior material for medical, garments, and technical textiles, production of silk has decreased with the advent of chemically produced fibers. Silk is nowadays partially supplanted by polyesters and viscose in the textile industry and by aramid and polyamide fibers in the defense industry.

4.1 Silk Market

The silk global market is projected to achieve a size of $25.4 billion by 2028, with a steady compound annual growth rate (CAGR) of 8% expected throughout the prediction period. In 2021, the German market held a leading position in the European silk market, attaining a market price of $1.4 billion by 2028. The UK market of is anticipated to encounter a rate of growth around 6.3% during the period from 2022 to 2028. Furthermore, the French market is expected to witness a CAGR of 8% during the same period (Knowledge Based Value, 2022).

4.2 Silk Waste

As it discussed previously regarding the textile waste, silk also produces a huge amount of waste at each step of its production. The farming, managing the cocoons of silkworms, and the production of raw silk generate substantial waste, involving by-products from cultivation of silkworms, unwinding of cocoon, spinning, and weaving of silk. Around 20% of cocoons are considered top-grade and can be entirely unwound without waste. The rest, classified as second and third grades, is not entirely treated in factories of silk. When adding together the silk surplus from weaving and spinning, the overall waste generated in every silk sector reaches 55%. However, the silk dyeing and finishing also contributes toward the wastage of silk (Sashina & Yakovleva, 2023).

4.3 Silk Value Chain

The "value chain"term introduced by Michael E. Porter (1985) refers to the entire order of activities necessary to take a service or product from its inception to its ultimate discarding. This comprises all phases of manufacturing, consumer delivery, and the subsequent product management after usage. It is reasonable to expect that the product value will likely rise as it progresses from one participant in the chain to another (Hellin & Meijer, 2006). The silk production is based on three main phases: (i) upstream, (ii) midstream, and (iii) downstream. Sericulture cultivation, silk rearing, and silk reeling are part of the upstream industry, while fabric weaving, dyeing, and printing constitute the midstream industry. In rural areas, these activities are typically performed manually, whereas in the industrial sector, large-scale machinery is employed for efficiency and productivity. The downstream industry is responsible for managing the designing, manufacturing, selling, and advertising of numerous silk items, like apparel, home décor, and bedding (Jatuphatwarodom et al., 2018; Patichol et al., 2014). Fig. 3 demonstrates the three silk value chain streams. Supplying raw materials such as silkworms, cocoons, and silk yarn for silk

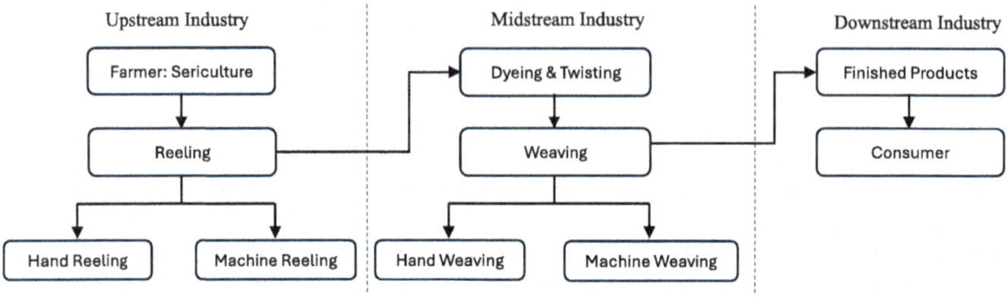

Fig. 3 The three silk value chain streams (Author's Creation)

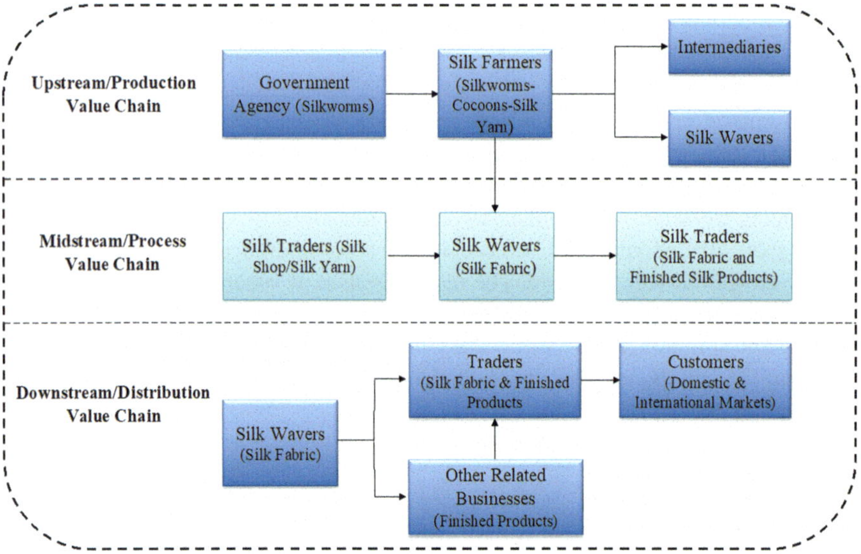

Fig. 4 Main stakeholders of silk value chain (Author's Creation). (Source: Rabia Hassan et al., 2023)

manufacturing falls within the value chain of upstream. The process of midstream involves the production of semi-processed products like fabric of silk, which is heavily induced by technical procedures.

Meanwhile, the downstream industry encompasses the distribution of semi-finished and finished products of silk. The procedure of decision-making at the level of factory encompasses the engagement of the three key silk stakeholders, namely farmers, weavers, and traders, as illustrated in Fig. 4. Individuals involved in the raising of silkworms are referred to as "silk farmers", and some of them also engage in the production of silk fabrics and yarn. Meanwhile, individuals specializing in weaving are recognized as silk weavers. However, traders act as intermediaries in the silk industry, involving in retailing silk products and managing silk-related trades.

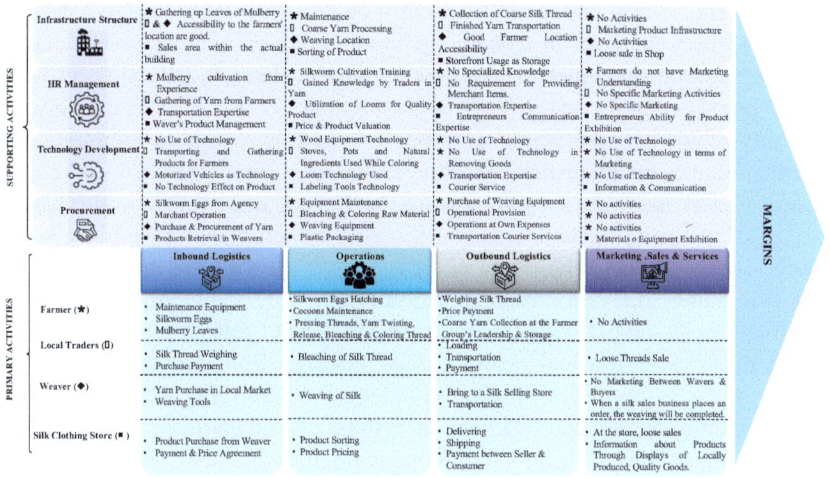

Fig. 5 Primary and supportive silk actors' activities (Author's Creation). (Source Rabia Hassan et al., 2023)

The major players involved in the silk industry are a farmer, trader, weaver, silk clothing stores, and consumers. Figure 5 depicts the primary and supportive activities of major players involved in the silk business.

4.3.1 Primary Activities

Farmers play an important part in the primary activities of the silk commodities sector, serving as manufactures or suppliers of coarse yarn derived from silkworms. Local traders are involved in the subsequent stage of the process, where their main activity consists of buying silk thread from farmers and then selling it to weavers/ craftsman. Moreover, weavers and craftsman are responsible for turning the threads into finished silk products. Ultimately, the primary participants in the silk commodities sector are the apparel retailers who also engage in weaving operations.

4.3.2 Supportive Activities

Concerning logistical infrastructure, farmers need sufficiently large fields to cultivate mulberry trees as a supporting activity. Coarse threads in outbound logistics are initially gathered in a box before being prepared for marketing. However, farmers do not handle marketing activities themselves; for yarn collection, they simply await local traders. In terms of internal logistics, farmers are proficient at using knives to cut mulberry leaves during the harvesting process. However, they lack specialized abilities for outbound logistics, as it primarily involves the silk thread collection, for which no specialized knowledge is required. Similarly, farmers do not possess specialized marketing skills, as they rely on local traders to handle the marketing aspect. In terms of technology, farmers rely on relatively simple equipment, such as shelves which are self-maintained made from bamboo and wood, bamboo-based materials, and equipment for spinning. While farmers are involved in supporting activities like procurement, the equipment they need is basic, including matches or knives, maintenance tools, and scales for weighing yarn.

Local traders are aided by infrastructure and accessibility, notably through the presence of adequate roads connecting them to farmers' sites. Regarding human resources, these traders rely on the expertise passed down from their parents and peers in activities like yarn bleaching. Additionally, they acquire marketing expertise through firsthand experiences. Local traders depend solely on motorized vehicles for the transferring and obtaining raw material from farmers. In the yarn bleaching process, they utilize simple technology, including pots, stoves, and locally available natural ingredients. When it comes to procurement, they operate independently, relying on their own capital for conducting business activities.

Weaving is conducted both at large industrial scale and at residences as weaving hubs, where they transform purchased threads into various clothing items. Weavers rely on their skills in operating motorized conveyance to efficiently gather threads from local traders. Furthermore, the process of weaving is facilitated by basic technologies, such as manual or automatic looms acquired from loom makers. Additionally, regarding procurement logistics, weavers receive support from silk clothing stores. In essence, the yarn acquired by weavers originates not only for purchases made through local traders but also through these silk clothing stores, highlighting the interconnected relationships between them.

Silk clothing stores use their own sales locations within their buildings for storage and sorting of goods purchased from weavers. They also employ cross-section containers for this purpose. For marketing, these stores organize sales on-site and showcase products at exhibitions. Silk clothing stores excel in human resources, effectively managing incoming products and pricing. They rely on communication skills for outbound logistics and marketing, utilizing smartphones for information gathering. In terms of technology, they use labeling tools for pricing clarity but rely on traditional freight forwarding for logistics.

5 Textile Waste Management Circular Strategies

When it comes to getting rid of their clothing, customers have a lot of alternatives. But the conventional method of just throwing away clothing after using it creates an accumulation of waste and pollution in the environment. This procedure not only overcrowds landfills but also endangers ecosystems by releasing toxic chemicals from degradation of textiles. In contrast, consumers can opt for more sustainable approaches for garment disposal. For example, they can donate their lightly worn clothes to charities or thrift stores, where others can benefit from them (Janigo & Wu, 2015). Another option is participating in clothing exchanges or sharing initiatives, which allow garments to find new homes and prolong their usefulness. Additionally, consumers can choose recycling programs that transform textiles into new products, reducing the need for new materials and lessening environmental harm (Stanescu, 2021). By embracing these alternative disposal methods, consumers can contribute toward the reduction of environmental impact due to garment waste. These practices not only alleviate the strain on landfills but also promote resource conservation and support a more circular fashion consumption model.

Today, numerous retail outlets, municipalities, collectors, and charitable organizations gather used garments, either for resale, donation to those in need or for recycling (Tuzcu & Kirezli, 2022). Even garments deemed non-reusable are repurposed by professional collectors into valuable items. Typically, discarded garments are either incinerated or deposited in landfills as solid waste (Wojnowska-Baryła et al., 2024). However, this solid waste can undergo sorting and be directed toward recycling processes. After undergoing recycling, textile waste can be repurposed into a range of items including rags, blankets, wipes, carpets, and insulation materials. Despite the possibilities, the technology for this recycling process remains limited, leading to a current recycling rate of only 25% of solid textile waste (Maity et al., 2020). Figure 6 illustrates the waste management strategies for the textile industry through which circularity can be attained in this domain.

5.1 *Reuse*

Reusing helps to decrease the need for producing new textiles (Chen et al., 2021). It refers to "slow fashion" or "durable fashion", which consists of robust materials made to last unlike fast fashion that is typically recycled instead of reused (Earley & Goldsworthy, 2015). As per the report of ThredUP (2019), reuse helps to slow down the fashion industry's fast increasing pattern. However, in Europe, the portion

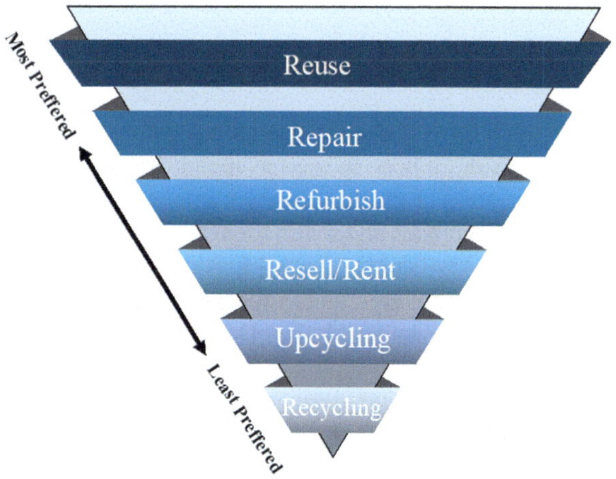

Fig. 6 Textile waste management strategies (Author's Creation). (Inspired by Bide et al., 2020)

of the solid textile waste that undergoes reuse is 8%, which is still less than the portion that is recycled (10%), both figures being notably lower than the portion of the solid textile waste that is ultimately disposed of in landfills (57%) (Keßler et al., 2021). However, it is noteworthy that reused textile is estimated at just 1%, according to the European Parliament (Nikolina, 2019).

Apart from the product's quality, the category of product is also important for its potential reuse. Baby or children's clothing, for example, is typically utilized for a brief period, making it less susceptible to wear and tear. Consequently, these types of products are frequently offered in thrift stores or contributed for further use. In 2020, over 80% of the top-selling garments via Vinted France resale platform originated through Petit Bateau, which is a French producer of children's and baby clothing (Statista, 2021). Apart from children's and baby clothing, luxury brands, designer pieces, and vintage items are also commonly available for reuse (ETC/WMGE, 2021).

According to a 2020 Internationale Nederlanden Groep survey, roughly 20% of consumers frequently purchase secondhand clothing, indicating that overall participation in the pre-owned market remains relatively limited. Particularly, it is the younger consumer demographic, specifically Generation Z and Millennials, who propel the growth of the secondhand market. Over 40% of individuals have purchased secondhand clothing, footwear, or accessories in 2020 (ThredUP, 2021). Online consumer-to-consumer (C2C) platforms provide a suitable method to effortlessly sell and buy old outfits. During the COVID-19 pandemic, these platforms experienced a significant increase in popularity as a means of selling apparel, putting them in a strong position to capitalize on the growth of the secondhand marketplace, which is projected to expand at a global CAGR of 15-20% in the coming years (Boston Consulting Group, 2020). Though the rise in consumer-to-consumer reuse aligns with circular economy principles, it shifts the reuse high-value segment away from collecting and sorting programs, leading to a notable impact on profitability.

5.2 Repair

Repairing textiles and clothing extends their usable lifespan. This step involves fixing buttons or zippers, as well as mending seams or the fabric structure itself. The tendency to repair garments varies among individuals, depending on their unique clothing preferences and sentimental attachment to the styles (McNeill et al., 2020). Consumers may face hurdles such as limited time and expertise for repairs, alongside the comparatively higher expenses associated with some repair techniques than to buying new cheaper clothes (Mclaren & Mclauchlan, 2015; Laitala & Klepp, 2018). Furthermore, people tend to be more inclined to repair expensive garments rather than cheaper textiles and fashion items (Diddi & Yan, 2019). Community-based initiatives for repair should be implemented as an effective platform to raise awareness regarding the importance of repairing in the circular economy. This can help individuals who are unfamiliar with repairing techniques of textile and introduce a social aspect to the specific material-focused process for repair (van der Velden, 2021). However, the textile industry needs to offer free repair services to customers through its own stores or by cooperating with local repair shops. These services should include offers like mending tears, replacing buttons or zippers, and meeting other common needs for repair.

5.3 *Refurbish*

In the textile industry, refurbish strategy can solve the issues caused by changing fashions and getting bored with clothes worn for a long time. It also solves the problem of clothes that no longer fit due to changes in body size. Refurbishment is not just changing the garment but also redesigning it, such as adding new details or changing the color to match current fashion trends (Winquist et al., 2023). It can be done by companies or consumers, especially when clothes are rented rather than purchased directly (Fischer & Pascucci, 2017). The manufacturer can refurbish the product to the same original design and return it to the customer (Kumar & Carolin, 2019). When the environmental consequences of refurbishing items such as event tents and beach flags were examined, the findings revealed a degradation in environmental effect ranging from 45% to 88% (Kaddoura et al., 2019). However, this situation is unique. Even if it is a small business for frequently purchased clothing, there are issues to be resolved, such as the need to verify information and the lack of guarantee of having to pay back and forth (Brauner et al., 2022). Addressing these refurbish challenges, including reuse and recycle is significant for amplifying circularity within the textile industry.

5.4 *Resell*

The fashion industry has seen a rise in the popularity of resale as a circular strategy, which effectively keeps garments out of landfills. Fashion businesses commonly endorse this strategy, using models like business-to-business (B2B) and business-to-consumer (B2C), to extend the apparel pieces' durability (Hedegård et al., 2020; de Aguiar Hugo et al., 2021). Fashion businesses are key players in driving the growth of the garment reselling market. However, individuals also take part in consumer-to-consumer (C2C) reselling of unwanted clothing items. Nonetheless, only a small fraction of users actively engage in consumer-to-consumer reselling (Weber et al., 2017). Reselling pre-owned textile items can draw in fresh customer demographics, boost customer loyalty, showcase corporate accountability, and generate extra income for manufacturers or brands. Several factors influence the effectiveness, including brand reputation, fashion trends, market development, and the maintenance of superior product quality. The resale market has amplified 21 times more quickly than traditional retail in the past 5 years (ThredUP, 2019). However, the retailers resold 101.2 million clothing items in 2020 from which only 560,000 items were secondhand (ThredUP, 2021).

5.5 *Upcycling*

Upcycling involves enriching the inherent value of materials through a design-driven approach, facilitated by a designer's innovative effort to reimagine disassembled and surplus garments and textiles' novel products. Upcycling transforms discarded textiles, replacing their former associations of waste and uselessness with a fresh significance and value, characterized by sustainability, superior quality, and timeless fashion (Eleonora Chiais, 2022). It is found that in textile industry, large manufacturing companies produce more leftover as compared to small factories. Figure 7 illustrates the average leftovers in garment manufacturing. Until recently, upcycling has primarily been employed in smaller, niche markets, often featuring one-of-a-kind items or serving as supplementary components in certain collections, rather than being applied to a large, industrial level (Moorhouse & Moorhouse, 2017).

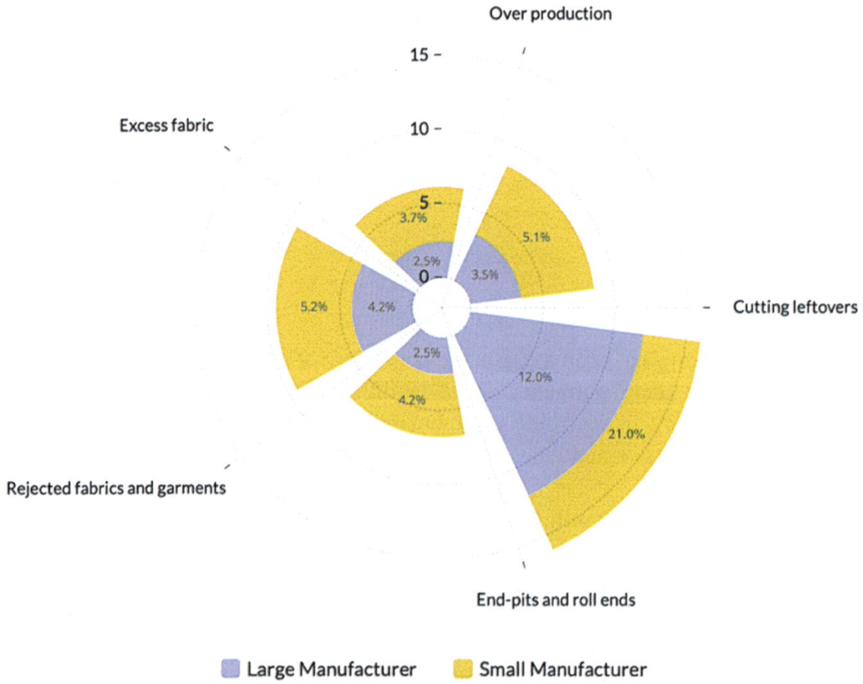

Fig. 7 Manufacturing garments' pre-consumer average fabric leftovers (Author's Creation). (Source: Aus et al., 2021)

The pre-consumer fabric leftovers can be upcycled by using various methods depending upon the leftovers, for instance, a dress can be done by using cutting leftovers. Designing a product comprising identical smaller parts is among the best strategies to limit cutting waste. This allows for the incorporation of these details into an existing marker, with the product being cut simultaneously.

The specifics will be gathered apart from the cutting table and forwarded to the sewing process. This approach provides a chance to reduce cutting waste by as much as 60%, varying the aspect added to the marker (Aus et al., 2021). The 80% leftovers of short rolls and end-pits can also be recycled, and the square shape of the pattern waste can use in the production of T-shirts, making it possible to implement a "zero waste" strategy. Textile waste in the form of longer roll ends, typically ranging from 3 to 49 meters, is the most prevalent type, and owing to the size of it, they are the most appropriate material for producing substantial amounts of upcycled clothing items. To enhance competence, roll ends exceeding 3 meters in length are directed to be segregated from the shorter ones. They are then either re-rolled and stored separately or immediately dispatched to produce upcycled garments.

In the upcycled system's design phase, it is essential to consider the scarcity of materials available for product development. This necessitates a significant degree of creative and innovative design thinking. To achieve this objective, it is vital to synchronize the principles within the supply network, encouraging interaction and cooperation among the stakeholders engaged in the supply chain. Sharing knowledge and skills demands substantial collaborative endeavors, yet it leads to a decrease in related expenses and enhances market performance.

5.6 *Recycling*

Recycling discarded textiles, whether waste from post-consumer or pre-consumer, typically begins with materials' sorting. After that, the mechanical, chemical, or bio-recycling methods take place. Recycling is simpler for textiles made from a single material compared to those composed of multiple materials. The potential uses for recycled fibers vary based on the textile waste quality; they can either be turned into new clothing fibers or used for composites and other less valuable products (Winquist et al., 2023).

5.6.1 Silk Recycling Methods

The silk waste can be recycled through three methods namely, bioenergy, mechanical, and chemical, as illustrates in the Fig. 8. The recycling of SW provides a valuable outcome as the nature of silk is biodegradable and biocompatible (Dhruval et al., 2020), with unique thermal conductivity, mechanical strength, and insulating electrical properties (Lakshmipriya & Imayathamizhan, 2021).

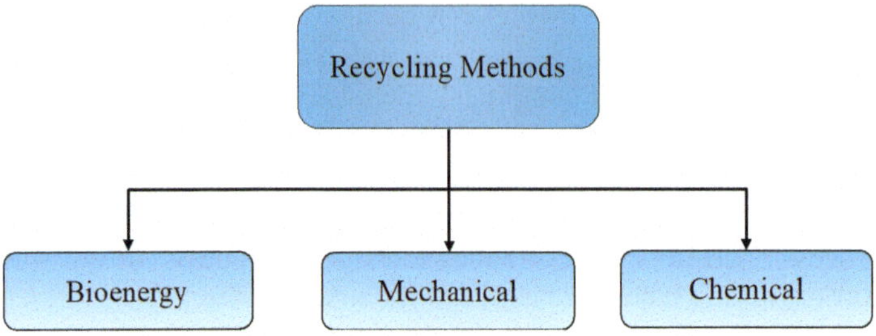

Fig. 8 Silk waste recycling methods

5.6.2 Bioenergy Conversion

Most textile wastes, involving SW, according to Leal Filho et al. (2019), are typically transformed into bioenergy by means of combustion. This process or recycling approach is highly direct and effective, alleviating the pressure of landfill disposal and potentially reducing volume by as much as 99%. Nevertheless, burning textile waste of one ton emits ten tons of CO_2 into the ambience (Zamani et al., 2015). Hence, it is certainly not an optimal solution for utilizing textile wastes like silk waste (SW). The principal element found in silk is protein, which can serve as a source of energy for poultry when used as feed. A feeding experiment was carried out with chicks as the subjects to investigate the impact of SW powder on their egg-laying over a period of 30 days. Findings revealed that feeding hens a diet incorporating 5% SW powder exhibited a 3.25% increase in egg production (Kim et al., 2002). There is potential for SW to undergo processing into biofuel, offering an alternative to fossil fuels.

5.6.3 Mechanical Recycling

This method is a frequently used technique for recycling SW, where the material is treated by melting and impregnation, altering the silk's molecular structure. The natural fiber can also be used for the development of the reinforced components which have several advantages like high strength, low cost, light weight, and have less impact on the environment. SW can be recycled into fiber assemblies, which can then be used as raw materials for fabric and yarn as illustrated in Fig. 9. Papori Hazarika et al. (2017) indicate that degummed SW combined with pineapple leaf fibers in different ratios can spun into yarns using a method of jute spinning. Subsequently, the fabric was produced with that blended yarn, which surpassed SW fabric in terms of feel, shine, and texture. This improved fabric can be used for women's apparel, shoulder shrugs, jackets, decorative textiles, home decor, and blinds for window. Koçak et al. (2008) used an extrusion molding technique to produce a composite material from silk waste fibers and high-density polyethylene. The composite mixture was then processed through an injection molding machine to create the fiber composite which results in high-strength material composite.

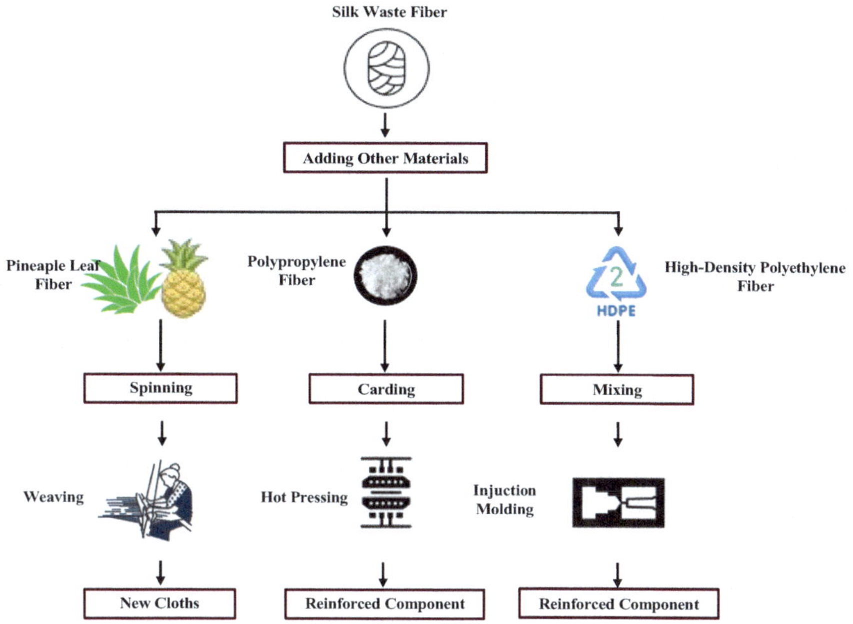

Fig. 9 Mechanical recycling methods (Author's Creation)

Similarly, Govindaraju and Jagannathan (2018) developed composite by using silk waste and polypropylene fibers; for this purpose, the authors first produced a mixture of fiber using these two waste. After that, the mixture of fiber was processed through a carding machine for producing fiber webs. These webs were subsequently compressed under heat to form composites using a molding press.

5.6.4 Chemical Recycling

The chemical recycling method is used for extracting silk protein from SW textiles to create regenerated silk fibroin. The silk fibroin extraction procedure from products of SW basically consists of three steps: bleaching or degumming, dissolving, and dialysis. Therefore, the initial phase is to eliminate sericin using methods such as alkali, acid, and soaping through high pressure and temperature, or techniques of enzymatic under specific conditions (S. Chen et al., 2019). However, for colored silk fabric waste, the initial step involves removing the coloring. The second phase involves dissolving the degummed silk, often accomplished using solutions like ionic liquids or formic acid-salt. During the final dialysis step, the inorganic ions in the regenerative silk fibroin mixture diffuse from within the dialysis bag into the surrounding water until osmotic pressure equilibrium is achieved on the bag's both sides. Once dialysis is finished, the purified regenerative silk fibroin mixture is obtained, which is a natural protein from animals, that can be modified by adding new groups onto them, altering the silk's structure of the molecule and new material production and can be used in various applications.

6 Silk Applications in Different Sectors

6.1 Dietary

Candy food products utilize the highest quality silk powder, silk hydrolyzed proteins, and amino silk acids. Silk proteins are utilized in specialized diets for diabetic and cardiac patients because of their digestibility, low levels of sugar and cholesterol, and ability to provide extra energy (Buhroo et al., 2018). Silk waste and cocoon dried powder are utilized as feed for fish and poultry, leading to improved growth rates and quality of hen's egg, and enhanced survival, fish eatable muscle growth, and rates of feed conversion (Mahanta et al., 2023).

6.2 Biomedical

Silk fiber has long been employed in surgical sutures because it avoids anarchic effects and dissolves once wounds heal. Further favorable medical uses include its potential as microtubes, which are biodegradable, for repairing the veins and as molded insertions for re-creating cartilage, bone, and teeth (Das & Natarajan, 2019). The strength and adhesive properties of sericin render silk fiber a viable option for biomedical applications requiring sealing and joining. In the bioengineering and biomedical realm, the incorporation of natural fibers with bio-resorbable polymers and biodegradable can fabricate bone and joints fixtures, offering pain relief for patients. Silk waste, abundantly accessible, can be converted into materials with varying diameters and lengths, resulting in improved performance and bio-stability. Silk sericin, a macromolecular protein extracted from silk fiber, demonstrates protection from ultraviolet (UV) radiations and oxidation, along with high moisture immersion and release capabilities. This protein can readily undergo cross-linking, copolymerization, and blending with additional macromolecular resources, particularly synthetic polymers, to fabricate enhanced properties in biocompatible medical and mechanical devices (Kapoor & Kundu, 2016).

6.3 Pharmaceutical

Silk is utilized in the treatment of cystitis, edema, impotence, epididymitis, therapy of adenosine augmentation, and cancer (Jatinder Singh Dhaliwal, 2019). A derivative of silk protein, known as serratiopeptidase, serves as an anti-tumefacient and anti-inflammatory agent. It is employed in the treatment of severe sinusitis, tonsillectomy, surgery, tonsillectomy, as well as dental procedures such as cleaning, tooth filling, and extractions (Spilsbury, 2018). Sericin has been found to inhibit tumor growth, to protect against ulcers, and to support various health benefits, including promoting cell proliferation to aid in wound-healing by forming a foundational matrix. Sericin is utilized in clinical diagnostics, in the production of crucial medical enzymes, and in the reduction of levels of blood glucose. Silk fibroin serves as an effective material for dressing due to its nontoxic nature of the tissues and is also employed in veterinary medicine (Suryawanshi et al., 2020).

6.4 Cosmetic

Silk fibroin peptides are incorporated into cosmetics because they provide a flexible, elastic, and shiny finish, along with excellent strong adhesion and spread ability (Tran Le Bao Ha et al., 2013). Silk sericin, known for its revitalizing, moisturizing, and UV-absorbing properties, is effective as an anti-irritant, skin moisturizer, sun protector, and anti-wrinkle agent, while also enhancing hair shape (Sehnal, 2008). Silk fiber protein extracts, including silk amino acids, hydrolyzed silk, raw silk fiber, and silk powder, are employed in the creation of personal care products, soaps, and cosmetics (Jaiswal et al., 2021). Silk powder is highly valued for its ability to soothe sunburns, thanks to its crystalline structure that reflects UV radiation. It acts as a demulcent, providing a protective barrier between the environment and skin. Finely micronized silk powder is often incorporated into massage oils and hair, whereas the more refined, water-soluble silk powder is utilized in various liquid cosmetic products.

6.5 House Building

In the construction sector, silk fibers are gaining attention due to their cost-effectiveness and superior insulation properties compared to existing materials. The mechanical characteristics of polypropylene and silk fiber composites have been found to surpass those of standard glass fiber-reinforced plastics (GFRP) for residential construction (Mann et al., 2023). The elasticity and tensile properties of the compound improved with a higher silk fiber content. Notably, the Izod impact strength of the silk–polypropylene (S–PP) composite matched that of GFRP, while also demonstrating better adhesion compatibility due to the interaction between the hydroxyl groups in polypropylene and the amino groups in silk. As a result, the incorporation of renewable composites of silk fiber-reinforced polymer is emerging as a crucial design consideration for industrial and residential construction materials.

7 Conclusion

In this chapter, the impacts on the environment for producing key textile fibers cotton, jute, and silk are discussed. This chapter places a strong emphasis on the procedures involved in transforming these three materials into fibers, with a particular focus on the environmental consequences at each stage of production. Additionally, it presents a detailed comparative analysis of natural fibers—cotton, jute, and silk—based on their environmental impact. The analysis specifically examines the agricultural, spinning, and weaving phases of yarn production. The findings reveal that silk fiber has the highest environmental impact in each phase when compared to cotton and jute. However, the dynamics of the market regarding the silk industry are also examined. Despite its profitability, the silk market silk produces a significant amount of waste, leading to loss of resources which are valuable and environmental risks.

Conversely, the exploration of strategies related to the circularity in the industry of textile discovers the prospective of obtaining effective and sustainable system. These strategies involve various methods comprising reusing, repairing, renting, reselling, refurbishing, upcycling, and recycling and can help to attain circularity and lessen waste and the industry's negative environmental impact. Furthermore, the revolutionary methods for silk waste recycling reveal how the industries can convert silk waste into valuable products like new clothing and reinforcement composite and how this will minimize the impact of textile waste and boost the economic value.

This chapter also explores silk's diverse applications across various sectors, highlighting its potential and versatility in promoting sustainability. Silk is one of the natural fibers through which the industries can fulfill the various requirements by using the waste and using it for different purposes in the cosmetic, construction, and medical industries.

It is imperious that silk industry has the capability of transforming toward a circular economy through which we can decrease the environmental effect of this industry and take advantage from the economic point of view. With this change, the silk industry can increase the economic sustainability and lower its adverse impact on the environment. However, to do this the collaboration among stakeholders, researchers, manufacturers, traders, consumers, and policymakers is essential. Similarly, developing and enforcing resilient policies is crucial for the textile industry's survival and success in an increasingly environmentally conscious world.

8 Future Recommendations

The characteristics of the silk material biodegradability provide the prospective to promote sustainability in the textile industry; to guarantee silk's sustainability, there is a need to educate consumers regarding its each phase. Individuals might help create a more sustainable fashion industry by getting more ecologically aware, by learning that silk is recyclable and biodegradable, and by making greener purchases. This informative effort might stress the environmental benefits of selecting silk against non-biodegradable alternatives and offer advice on appropriate disposal methods like recycling or composting. This is a crucial element for developing environmentally responsive approaches in the textile industry and increasing customer awareness of the usage of environmentally responsive assets like silk.

However, there is a need for comprehensive value chain framework incorporating all actors involved in the silk industry to achieve circularity. This practice would make it easier to retrieve silk goods once their practical uses are gone. Reusing resources for silk manufacturing may be facilitated by implementing sustainable techniques throughout the process.

Acknowledgments This study was carried out within the MICS (Made in Italy—Circular and Sustainable) extended partnership and received funding from the European Union Next-Generation EU (PIANO NAZIONALE DI RIPRESA E RESILIENZA (PNRR)—MISSIONE 4 COMPONENTE 2, INVESTIMENTO 1.3—D.D. 1551.11-10-2022, PE00000004). The content of this chapter represents only the authors' perspectives and does not reflect the views of the European Union or the European Commission, which are not responsible for any of the information presented.

Conflict of Interest The authors have no plausible variance of interest.

References

Ahmed, A. T. M. F., Islam, M. Z., Mahmud, M. S., Sarker, M. E., & Islam, M. R. (2022). Hemp as a potential raw material toward a sustainable world: A review. *Heliyon, 8*(1), e08753. https://doi.org/10.1016/j.heliyon.2022.e08753

Andadari, L., Yuniati, D., Supriyanto, B., et al. (2022). Lens on tropical sericulture development in Indonesia: Recent status and future directions for industry and social forestry. *Insects, 13*(10). https://doi.org/10.3390/insects13100913

Aus, R., Moora, H., Vihma, M., Unt, R., Kiisa, M., & Kapur, S. (2021). Designing for circular fashion: integrating upcycling into conventional garment manufacturing processes. *Fashion and Textiles, 8*(1). https://doi.org/10.1186/s40691-021-00262-9

Babu, K. M. (2018). *Silk: Processing, properties and applications*. Elsevier.

Bahl, R., Panwar, T., Padhye, R., & Nayak, R. (2023). Are Australian consumers ready to wear recycled clothing to practice sustainable consumption? *Sustainability, 15*(21), 15451. https://doi.org/10.3390/su152115451

Barcelos, S. M. B. D., Salvador, R., da Graça Guedes, M., & de Francisco, A. C. (2020). Opportunities for improving the environmental profile of silk cocoon production under Brazilian conditions. *Sustainability (Switzerland), 12*(8). https://doi.org/10.3390/SU12083214

Tom Bide, Teresa Brown, Ellie Evans, Naomi Idoine, Carolin Kresse, Evi Petavratzi, Lucia Mancini, & Beatriz Vidal-Legaz. (2020). *Optimizing quality of information in Raw Material data collection across Europe - ORAMA*. http://www.orama-h2020.eu

Boston Consulting Group. (2020). *The consumers behind fashion's growing secondhand market*. https://www.bcg.com/press/2november2020-the-consumers-behind-fashions-growing-secondhand-market

Brauner, P., Vervier, L., Ziefle, M., Sachtleben, M., Schlichter, S., & Gries, T. (2022). Clothing-as-a-service? A research agenda towards a sustainable and socially accepted circular economy of clothing. *Human Factors for Apparel and Textile Engineering, 32*. https://doi.org/10.54941/ahfe1001551

Buhroo, Z. I., Bhat, M. A., Malik, M. A., Kamili, A. S., Ganai, N. A., & Khan, I. L. (2018). Trends in development and utilization of sericulture resources for diversification and value addition. *International Journal of Entomological Research, 6*(1), 27–47. https://doi.org/10.33687/entomol.006.01.2069

Cao, H., Cobb, K., Yatvitskiy, M., Wolfe, M., & Shen, H. (2022). Textile and product development from end-of-use cotton apparel: A study to reclaim value from waste. *Sustainability (Switzerland), 14*(14). https://doi.org/10.3390/su14148553

Cappellozza, S., Casartelli, M., Sandrelli, F., Saviane, A., & Tettamanti, G. (2022). Silkworm and silk: Traditional and innovative applications. *Insects, 13*(11). https://doi.org/10.3390/insects13111016

Chen, S., Liu, M., Huang, H., Cheng, L., & Zhao, H. P. (2019). Mechanical properties of Bombyx mori silkworm silk fibre and its corresponding silk fibroin filament: A comparative study. *Materials and Design, 181*, 108077. https://doi.org/10.1016/j.matdes.2019.108077

Chen, X., Memon, H. A., Wang, Y., Marriam, I., & Tebyetekerwa, M. (2021). Circular economy and sustainability of the clothing and textile industry. *Materials Circular Economy, 3*(1). https://doi.org/10.1007/s42824-021-00026-2

Chiais, E. (2022). From"out of fashion" to upcycling. How art and the fashion industry have resemantized the word "recycling". *ZoneModa Journal, 12*(1), 135–144.

Das, S., & Natarajan, G. (2019). Silk fiber composites in biomedical applications. In *Materials for biomedical engineering* (pp. 309–338).

de Aguiar Hugo, A., de Nadae, J., & da Silva Lima, R. (2021). Can fashion be circular? A literature review on circular economy barriers, drivers, and practices in the fashion industry's productive chain. *Sustainability, 13*(21), 12246.

Dhruval, S. R., Pai, N., Dhanwant, S. S., Hussein, B., Nayak, S., Rao, C. V., Kumar, A., & Janakaraj, M. (2020). Rapid synthesis of antimicrobial Fe/Cu alloy nanoparticles using Waste Silkworm Cocoon extract for cement mortar applications. *Advances in Natural Sciences: Nanoscience and Nanotechnology, 11*(2). https://doi.org/10.1088/2043-6254/ab8790

Diddi, S., & Yan, R. N. (2019). Consumer perceptions related to clothing repair and community mending events: A circular economy perspective. *Sustainability (Switzerland), 11*(19). https://doi.org/10.3390/su11195306

Dobilaite, V., Mileriene, G., Juciene, M., & Saceviciene, V. (2017). Investigation of current state of pre-consumer textile waste generated at Lithuanian enterprises. *International Journal of Clothing Science and Technology, 29*(4), 491–503. https://doi.org/10.1108/IJCST-08-2016-0097

Dragomir, V. D., & Dumitru, M. (2022). Practical solutions for circular business models in the fashion industry. *Cleaner Logistics and Supply Chain, 4*, 100040. https://doi.org/10.1016/j.clscn.2022.100040

Earley, R., & Goldsworthy, K. (2015). Designing for fast and slow circular fashion systems: Exploring strategies for multiple and extended product cycles. *Product lifetimes and the environment. Nottingham*. https://www.researchgate.net/publication/279190953

Ellen MacArthur Foundation. (2013a). *Circular economy towards the economic and business rationale for an accelerated transition*.

Ellen MacArthur Foundation. (2013b). *Towards the circular economy*. https://www.aquafil.com/assets/uploads/ellen-macarthur-foundation.pdf

Ellen MacArthur Foundation. (2017). A new textiles economy: Redesigning fashion's future. In *Ellen MacArthur Foundation*.

Esbeih, K. N., Molina-Moreno, V., Núñez-Cacho, P., & Silva-Santos, B. (2021). Transition to the circular economy in the fashion industry: The case of the inditex family business. *Sustainability (Switzerland), 13*(18). https://doi.org/10.3390/su131810202

ETC/WMGE. (2021). *Business models in a circular economy*. https://www.eionet.europa.eu/etcs/etc-wmge/products/etc-wmge-reports/business-models-in-a-circular-economy

Fernandes, L., Rosado da Cruz, A. M., Cruz, E. F., & Lopes, S. I. (2023). A Review on Adopting Blockchain and IoT Technologies for Fostering the Circular Economy in the Electrical and Electronic Equipment Value Chain. *Sustainability (Switzerland), 15*(5). https://doi.org/10.3390/su15054574

Fidan, F. Ş., Aydoğan, E. K., & Uzal, N. (2021). *A comparative life cycle assessment of conventional and organic cotton in denim fabric*.

Fischer, A., & Pascucci, S. (2017). Institutional incentives in circular economy transition: The case of material use in the Dutch textile industry. *Journal of Cleaner Production, 155*, 17–32. https://doi.org/10.1016/j.jclepro.2016.12.038

Gjurašić, M., & Đurović, T. (2022). Development of sericulture in the Eastern Adriatic during the Austrian Administration. *Athens Journal of History, 9*(1), 9–52. https://doi.org/10.30958/ajhis.9-1-1

Gonzalez, V., Lou, X., & Chi, T. (2023). Evaluating environmental impact of natural and synthetic fibers: A life cycle assessment approach. *Sustainability (Switzerland), 15*(9). https://doi.org/10.3390/su15097670

Govindaraju, R., & Jagannathan, S. (2018). Optimization of mechanical properties of silk fiber-reinforced polypropylene composite using Box–Behnken experimental design. *Journal of Industrial Textiles, 47*(5), 602–621. https://doi.org/10.1177/1528083716667257

Ha, T. L. B., Quan, T. M., & Vu, D. N. (2013). Naturally derived biomaterials: Preparation and application. In *Regenerative medicine and tissue engineering*. IntechOpen.

Haq, U. N., & Alam, S. M. R. (2023). Implementing circular economy principles in the apparel production process: Reusing pre-consumer waste for sustainability of environment and economy. *Cleaner Waste Systems, 6*, 100108. https://doi.org/10.1016/j.clwas.2023.100108

Hassan, R., Terzi, S., & Rosa, P. (2023). *Silk industry value chain and circular economy*.

Hazarika, P., Hazarika, D., Kalita, B., Gogoi, N., Jose, S., & Basu, G. (2017). Development of apparels from silk waste and pineapple leaf fiber. *Journal of Natural Fibers, 15*(3), 416–424.

Hedegård, L., Gustafsson, E., & Paras, M. K. (2020). Management of sustainable fashion retail based on reuse – A struggle with multiple logics. *International Review of Retail, Distribution and Consumer Research, 30*(3), 311–330. https://doi.org/10.1080/09593969.2019.1667855

Hellin, J., & Meijer, M. (2006). Guidelines for value chain analysis. *Analysis*.

Huygens, D., Foschi, J., Caro, D., Caldeira, C., Faraca, G., Foster, G., Solis, M., Marschinski, R., Napolano, L., Fruergaard Astrup, T., & Tonini, D. (2023). Techno-scientific assessment of the management options for used and waste textiles in the European Union. *European Commission, Joint Research Centre*. https://doi.org/10.2760/6292.

Ingram, M. (2002). Producing the natural fiber naturally: Technological change and the US organic cotton industry. *Agriculture and Human Values, 19*, 325.

Jaiswal, K. K., Banerjee, I., & Mayookha, V. P. (2021). Recent trends in the development and diversification of sericulture natural products for innovative and sustainable applications. In *Bioresource technology reports* (Vol. 13, p. 100614). Elsevier Ltd. https://doi.org/10.1016/j.biteb.2020.100614

Janigo, K. A., & Wu, J. (2015). Collaborative redesign of used clothes as a sustainable fashion solution and potential business opportunity. *The Journal of Design, Creative Process & the Fashion Industry, 7*(1).

Jatinder Singh Dhaliwal. (2019). Natural fibers: Applications. In *Generation, development and modifications of natural fibers*.

Jatuphatwarodom, N., Jones, D. F., & Ouelhadj, D. (2018). A mixed-model multi-objective analysis of strategic supply chain decision support in the Thai silk industry. *Annals of Operations Research, 267*(1–2), 221–247. https://doi.org/10.1007/s10479-018-2774-6

Kaddoura, M., Kambanou, M. L., Tillman, A. M., & Sakao, T. (2019). Is prolonging the lifetime of passive durable products a low-hanging fruit of a circular economy? A multiple case study. *Sustainability (Switzerland), 11*(18). https://doi.org/10.3390/su11184819

Kapoor, S., & Kundu, S. C. (2016). Silk protein-based hydrogels: Promising advanced materials for biomedical applications. *Acta Biomaterialia, 31*, 17–32. https://doi.org/10.1016/j.actbio.2015.11.034

Keßler, L., Matlin, S. A., & Kümmerer, K. (2021). The contribution of material circularity to sustainability – Recycling and reuse of textiles. *Current Opinion in Green and Sustainable Chemistry, 32*, 100535. https://doi.org/10.1016/j.cogsc.2021.100535

Kim, D., Kim, K., Kim, R., Chen, W., & Mitsuo, U. (2002). *Reuse of silk waste as poultry feed.*

Kirchherr, J., Reike, D., & Hekkert, M. (2017). Conceptualizing the circular economy: An analysis of 114 definitions. *Resources, Conservation and Recycling, 127*, 221–232. https://doi.org/10.1016/j.resconrec.2017.09.005

Kliś, S. (2023). Production waste of the textile and clothing industry in sustaindevelopment context. *Scientific Papers of Silesian University of Technology Organization and Management Series, 2023*(168). https://doi.org/10.29119/1641-3466.2023.168.14

Knowledge Based Value. (2022). *Market search report.* https://www.kbvresearch.com/silk-market/

Koçak, D., Tasdemir, M., Usta, I., Merdan, N., & Akalin, M. (2008). Mechanical, thermal, and microstructure analysis of silk- and cotton-waste-fiber-reinforced high-density polyethylene composites. *Polymer Plastics Technology and Engineering, 47*(5), 502–507. https://doi.org/10.1080/03602550801977919

Kumar, S., & Carolin, F. (2019). Future for circular economy. In *Circular economy in textiles and apparel* (pp. 207–217). The Textile Institute Book Series.

La Rosa, A. D., & Grammatikos, S. A. (2019). Comparative life cycle assessment of cotton and other natural fibers for textile applications. *Fibers, 7*(12). https://doi.org/10.3390/fib7120101

Laitala, K., & Klepp, I. G. (2018). Care and production of clothing in Norwegian Homes: Environmental implications of mending and making practices. *Sustainability (Switzerland), 10*(8). https://doi.org/10.3390/su10082899

Lakshmipriya, S., & Imayathamizhan, N. M. (2021). Development of silk/polycaprolactone biocomposite for internal bone plate application. *Journal of Industrial Textiles, 51*(1), 3–23. https://doi.org/10.1177/1528083719886052

Leal Filho, W., Ellams, D., Han, S., Tyler, D., Boiten, V. J., Paco, A., Moora, H., & Balogun, A. L. (2019). A review of the socio-economic advantages of textile recycling. *Journal of Cleaner Production, 218*, 10–20. https://doi.org/10.1016/j.jclepro.2019.01.210

Mahanta, D. K., Komal, J., Samal, I., Bhoi, T. K., Dubey, V. K., Pradhan, K., Nekkanti, A., Gouda, M. N. R., Saini, V., Negi, N., Bhateja, S., Jat, H. K., & Jeengar, D. (2023). Nutritional aspects and dietary benefits of "Silkworms": Current scenario and future outlook. *Frontiers in Nutrition, 10*, 1121508. https://doi.org/10.3389/fnut.2023.1121508

Maity, S., Mondal, M. K., Pandit, P., & Singha, K. (2020). Future mobilizations and paths of waste – Towards best solution. In *Recycling from waste in fashion and textiles: A sustainable and circular economic approach* (pp. 321–340).

Malinverno, N., Schmutz, M., Nowack, B., & Som, C. (2023). Identifying the needs for a circular workwear textile management – A material flow analysis of workwear textile waste within Swiss Companies. *Resources, Conservation and Recycling, 189*, 106728. https://doi.org/10.1016/j.resconrec.2022.106728

Mann, G. S., Azum, N., Khan, A., Rub, M. A., Hassan, M. I., Fatima, K., & Asiri, A. M. (2023). Green composites based on animal fiber and their applications for a sustainable future. *Polymers, 15*(3). https://doi.org/10.3390/polym15030601

Marcuello, C., Chabbert, B., Berzin, F., Bercu, N. B., Molinari, M., & Aguié-Béghin, V. (2023). Influence of surface chemistry of fiber and lignocellulosic materials on adhesion properties with polybutylene succinate at nanoscale. *Materials, 16*(6). https://doi.org/10.3390/ma16062440

McKinsey & Company. (2022). *Scaling textile recycling in Europe-turning waste into value.*

Mclaren, A., & Mclauchlan, S. (2015). *Crafting sustainable repairs: Practice based approaches to extending the life of clothes* (Vol. 17). Laitala & Boks.

McNeill, L. S., Hamlin, R. P., McQueen, R. H., Degenstein, L., Garrett, T. C., Dunn, L., & Wakes, S. (2020). Fashion sensitive young consumers and fashion garment repair: Emotional connections to garments as a sustainability strategy. *International Journal of Consumer Studies, 44*(4), 361–368. https://doi.org/10.1111/ijcs.12572

Moazzem, S., Crossin, E., Daver, F., & Wang, L. (2021). Assessing environmental impact reduction opportunities through life cycle assessment of apparel products. *Sustainable Production and Consumption, 28*, 663–674. https://doi.org/10.1016/j.spc.2021.06.015

Moorhouse, D., & Moorhouse, D. (2017). Sustainable design: Circular economy in fashion and textiles. *Design Journal, 20*(sup1), S1948–S1959. https://doi.org/10.1080/14606925.2017.1352713

Munasinghe, P., Druckman, A., & Dissanayake, D. G. K. (2021). A systematic review of the life cycle inventory of clothing. *Journal of Cleaner Production, 320*, 128852. https://doi.org/10.1016/j.jclepro.2021.128852

Nikolina, S. (2019). Environmental impact of textile and clothes industry.

Panthee, S., Paudel, A., Hamamoto, H., & Sekimizu, K. (2017). Advantages of the silkworm as an animal model for developing novel antimicrobial agents. *Frontiers in Microbiology, 8*, 373. https://doi.org/10.3389/fmicb.2017.00373

Papamichael, I., Chatziparaskeva, G., Voukkali, I., Navarro Pedreno, J., Jeguirim, M., & Zorpas, A. A. (2023). The perception of circular economy in the framework of fashion industry. *Waste Management and Research, 41*(2), 251–263. https://doi.org/10.1177/0734242X221126435

Patichol, P., Wongsurawat, W., & Johri, L. M. (2014). Upgrade strategies in the Thai silk industry: Balancing value promotion and cultural heritage. *Journal of Fashion Marketing and Management, 18*(1), 20–35. https://doi.org/10.1108/JFMM-09-2011-0059

Porter, M. E. (1985). *Competitive advantage*.

Riemens, J., Lemieux, A. A., Lamouri, S., & Garnier, L. (2021). A delphi-régnier study addressing the challenges of textile recycling in europe for the fashion and apparel industry. *Sustainability (Switzerland), 13*(21). https://doi.org/10.3390/su132111700

Sashina, E. S., & Yakovleva, O. I. (2023). The current state and prospects of recycling silk industry waste into nonwoven materials. *Fibers, 11*(6), 56. https://doi.org/10.3390/fib11060056

Schmutz, M., & Som, C. (2022). Identifying the potential for circularity of industrial textile waste generated within Swiss companies. *Resources, Conservation and Recycling, 182*, 106132. https://doi.org/10.1016/j.resconrec.2021.106132

Sehnal, F. (2008). Prospects of the practical use of silk sericins. *Entomological Research, 38*(1), S1–S8. https://doi.org/10.1111/j.1748-5967.2008.00168.x

Şener Fidan, F., Kızılkaya Aydoğan, E., & Uzal, N. (2023). The impact of organic cotton use and consumer habits in the sustainability of jean production using the LCA approach. *Environmental Science and Pollution Research, 30*(4), 8853–8867. https://doi.org/10.1007/s11356-022-22872-6

Spilsbury, D. K. (2018). Restoring credibility to natural materials. In *Designing with natural materials* (pp. 133–163).

Stanescu, M. D. (2021). State of the art of post-consumer textile waste upcycling to reach the zero waste milestone. *Environmental Science and Pollution Research, 28*, 14253–14270. https://doi.org/10.1007/s11356-021-12416-9/Published

Statista. (2021). *Best selling second hand fashion brands on the* Vinted.fr *platform in France in 2020*. Statista. https://www.statista.com/statistics/1147808/best-selling-brands-second-hand-products-vinted-france/

Suryawanshi, R., Kanoujia, J., Parashar, P., & Saraf, S. (2020). Sericin: A versatile protein biopolymer with therapeutic significance. *Current Pharmaceutical Design, 26*(42), 5414–5429.

Textile Exchange. (2022a). *Organic cotton market report*.

Textile Exchange. (2022b). *Preferred fiber & materials: Market report 2022* (Issue October).

ThredUP. (2019). *2019 resale report*.

ThredUP. (2021). *2021 resale report*.

Todor, M. P., Bulei, C., Kiss, I., & Alexa, V. (2019). Recycling of textile wastes into textile composites based on natural fibres: The valorisation potential. *IOP Conference Series: Materials Science and Engineering, 477*(1). https://doi.org/10.1088/1757-899X/477/1/012004

Tuzcu, M., & Kirezli, O. (2022). An exploratory study on the role of municipalities in the circulation of second-hand textile products. In *Circular economy: Multidisciplinary approaches from Turkey* (pp. 147–162).

van der Velden, M. (2021). 'Fixing the world one thing at a time': Community repair and a sustainable circular economy. *Journal of Cleaner Production, 304*, 127151. https://doi.org/10.1016/j.jclepro.2021.127151

Weber, S., Lynes, J., & Young, S. B. (2017). Fashion interest as a driver for consumer textile waste management: reuse, recycle or disposal. *International Journal of Consumer Studies, 41*(2), 207–215. https://doi.org/10.1111/ijcs.12328

Winquist, E., Horn, S., Tuovila, H., Lavikko, S., Sorvari, J., Joutsjoki, V., Karhu, M., Slotte, P., Kautto, P., Kivikytö-Reponen, P., & Ilvesniemi, H. (2023). *R-strategies in circular economy: Textile, battery, and agri-food value chains* (Vol. 57).

Wojnowska-Baryła, I., Bernat, K., Zaborowska, M., & Kulikowska, D. (2024). The growing problem of textile waste generation – The current state of textile waste management. *Energies, 17*(7). https://doi.org/10.3390/en17071528

Xu, B., Chen, Q., Fu, B., Zheng, R., & Fan, J. (2022). Current situation and construction of recycling system in china for post-consumer textile waste. *Sustainability (Switzerland), 14*(24). https://doi.org/10.3390/su142416635

Zamani, B., Svanström, M., Peters, G., & Rydberg, T. (2015). A carbon footprint of textile recycling: A case study in Sweden. *Journal of Industrial Ecology, 19*(4), 676–687. https://doi.org/10.1111/jiec.12208

A Taxonomy of Sustainable and Circular Apparel Business Models

Elisa Arrigo and Flavio Gnecchi

1 Introduction

The continuous growth of the apparel industry produces a heavy consumption of natural resources, such as the use of energy and water in production, as well as the unauthorized use of hazardous chemicals for tanning and dyeing (Muthu, 2024). Between 2024 and 2028, the worldwide apparel market is expected to generate US$0.2 trillion in revenue, with an expansion of 11.17 percent (Statista, 2023). Moreover, the severe weather events of 2023 have increased public authorities' scrutiny of fashion supply chains and raised the possibility of new regulations on several fronts (State of Fashion, 2024).

New sustainability regulations in the European Union (EU) and the United States are going on, related to limiting the greenhouse gas emissions and waste of textile and apparel industries, while building business models that protect and preserve the natural resources. These facts, combined with customers' protests, have driven apparel companies toward the development of sustainability policies (Jung & Jin, 2016). Additionally, a circular economy approach (Geissdoerfer et al., 2020) has been adopted to move beyond the linear and unsustainable model of "take, make, and waste". Some apparel companies began to use more sustainable materials with lower carbon footprint, such as organic cotton or recycled polyester, since up to 80 percent of a product's environmental impact is defined in the design phase. For instance, the EU's flagship Eco-design for Sustainable Products Regulation (ESPR) dictates product level's ecological design and circularity standards, complemented by digital product passports. This involves constraints on recyclability, durability,

E. Arrigo (✉) · F. Gnecchi
Department of Economics, Management and Statistics (DEMS),
University of Milano-Bicocca, Milan, Italy
e-mail: elisa.arrigo@unimib.it; flavio.gnecchi@unimib.it

S. S. Muthu (ed.), *Circular Business Models in the Apparel Industry*,
Environmental Footprints and Eco-design of Products and Processes,
https://doi.org/10.1007/978-3-031-80193-8_3

reusability, repairability, and use of hazardous substances and is set to come into full effect by 2025 in the EU (State of Fashion, 2024).

It is not simple for apparel companies, however, to invest in sustainable product design to limit the environmental damage of textile manufacturing. Thus, some companies started moving toward the circular economy steps associated with use-oriented services with principles such as take-back, reuse, and recycling as integral parts of the value chain (Abbate et al., 2023; Bocken et al., 2014; Henry et al., 2020). In fact, from the consumption perspective, new circular business models have developed as hopeful strategies to reduce the ecological apparel waste of the apparel market (Abdelmeguid et al., 2024). This study examines the main and different circular business models related to the fashion consumption field, i.e., collaborative consumption models that are spreading in the apparel market.

The chapter is organized as follows: after the Introduction, Sect. 2 presents the circular economy diffusion in the apparel market; Sect. 3 describes the taxonomy of the various circular business consumption models with specific apparel brands' cases and market data. Finally, Sect. 4 concludes with future research directions.

2 Circular Fashion in the Apparel Market

The apparel industry is connoted by an inherent tension due to a rising low-cost consumption and an environmentally unsustainable production system that mostly employs nonrenewable resources and intensive water use (Abbate et al., 2023; Tunn et al., 2019; Todeschini et al., 2017). Over the past decade, on the one hand, consumer preferences have increasingly shifted toward purchasing larger quantities of garments at lower prices (Statista, 2023). In contrast, on the other hand, a growing component of fashion customers has become aware of climate change and social and environmental issues caused by the apparel industry and have changed their buying habits looking for sustainability. Dissanayake and Weerasinghe (2021) distinguished four approaches of fashion circularity during the product's life cycle that apparel companies can adopt to become more sustainable. The circularity approaches are resource efficiency, circular design, product life extension, and end-of-life circularity.

Resource efficiency is addressed to limit the resources' use during the manufacturing stage to decrease the created pollution (Kumar et al., 2022) and requires the use of sustainable and bio-based inputs that reduce the waste, adopting best practices, facilitating the circularity. To reach a relevant decrease of resources' consumption and waste limitation (Tunn et al., 2019), brands look for more sustainable materials and often need to find new suppliers or join strategic alliances. For example, in 2013, Kering created the Material Innovation Lab, devoted to discovering eco-friendly materials and textiles, to provide Kering businesses' creative teams a deeper grasp of eco-friendly materials. This is a real driver of change within a very complex supply chain (https://www.kering.com/).

Circular design plays a special role in developing durable products, with limited environmental impact. Better design and reuse can also decrease the cost of raw materials and landfill. Design for circularity implies creating products that can be used many times since their materials are designed to be disassembled and reused (Bocken et al., 2016). The sportswear company Decathlon sells "Ecodesign products" with longer durability since each product can be projected to last more. In fact, products are tested within the design centers and R&D lab facilities in Europe to manufacture goods that can be used, reused, repaired, and recycled over time (https://www.decathlon.com).

Product life extension extends the product life cycle to retain the garment in use for a long time through operational practices (Geissdoerfer et al., 2020). This is possible not only by making available repairing services for customers but also through leasing or rental models that extend the durability and longevity of fashion products (Feng et al., 2020; Arrigo, 2022; Liu et al., 2023). The wearing apparel company Patagonia, known for its high-quality outdoor apparel and commitment to environmental sustainability efforts, was among the first to offer customers a repair service for their garments. Furthermore, through video tutorials, the company also explains to customers how to repair the purchased items, thus educating them to adopt circular and sustainable practices (https://www.patagonia.com).

Finally, *end-of-life circularity* means to keep garments out of landfill, exploiting the value of their materials in different ways (Colucci & Vecchi, 2024). In fact, garments can be sold in the secondhand fashion market, shared through rental or swapping, or remanufactured through disassembling, renovating, and assembling them into a new item (Dissanayake & Weerasinghe, 2021). Swedish recycling company Renewcell is building the world's first large-scale fiber-to-fiber recycling factory, also in collaboration with global brands such as H&M, Inditex, and Levi's. Renewcell's products are made of 100 percent textile waste, thanks to a special technology that enables the fashion industry to move from a linear to a circular value chain (https://www.renewcell.com/en/).

As highlighted in the above brief descriptions, the first two circular fashion approaches, namely *resource efficiency* and *circular design*, are more related to the manufacturing process in a fashion supply chain, while *product life extension* and *end-of-life circularity* are associated with the consumption stage. These latter circular consumption models will be the focus of the next section.

3 Sustainable and Circular Apparel Business Models

A clear taxonomy of collaborative fashion consumption models has been proposed in a recent systematic literature review (Arrigo, 2021), as a result of a codification and categorization of aspects identified in previous academic articles. Collaborative fashion consumption business models have been categorized based on:

- The type of organizer: the exchanges can be only between customers (C2C) or between a business/entrepreneur and customers (B2C) (Schor & Fitzmaurice, 2015; Iran & Schrader, 2017).
- The form of remuneration (monetary or non-monetary) that can consist in money transfer or, in other cases, in shared products (Henninger et al., 2019).
- The presence or absence of ownership transfer (Park & Armstrong, 2017).

The results of the carried out systematic literature review identified sustainable and circular fashion business models (hand-me-downs, swapping, commercial fashion sharing platforms and subscription services, rental, and secondhand resale) (Arrigo, 2021) that will be examined in the current fashion landscape, illustrating successful business cases.

3.1 Hand-me-Downs

"Passing on" and/or "hand-me-downs" are considered the oldest forms of collaborative consumption as products, typically apparel or other personal possessions, are forever transferred from one user to another of family, friends, and peers (Ruckelshausen & Kamleitner, 2024). Thus, they involve customer-to-customer (C2C) exchanges, with nonmonetary remuneration and product transferred ownership (Arrigo, 2021). It has been shown that a similar product sharing within families is directed to foster strong family relationships and to establish social affinities (Holmes, 2018; Decoopman, Gentina & Fosse-Gomez, 2010). These circular practices are often different across countries. For instance, while women in France exchange clothing within the family to reclaim their feminine identity, in Japan mothers do not want to share clothing with daughters to maintain their position of superior authority (Gentina et al., 2017). Hand-me-downs are habitually used among young people within groups of friends as teenagers often share clothes to cope with everyday problems, meet new people, and experience and differentiate themselves from their peer group (Gentina, 2014). In fact, being accepted by the peers plays a fundamental role in influencing the decision-making of teenagers (Gentina & Fosse-Gomez, 2010).

3.2 Swapping

Swapping represents a permanent change of ownership and circulation of unused or undesired fashion items in order to maximize their utility and longevity (Park & Armstrong, 2017). It is often connoted by a C2C exchange and a nonmonetary remuneration since clients do not receive money but apparel (Henninger et al., 2019). Swapping model supports an increased use of apparel items as an alternative of acquiring new items with important social implications for customers (Lang et al., 2020; McNeill & Venter, 2019). In fact, customers' intentions to exchange clothes derive positively from their "need for uniqueness" and "fashion leadership" that are superior to the barriers associated with materialism or contamination risks (Clube & Tennant, 2020). Additionally, customers are fascinated to participate in swapping events for the opportunity to wear a new dress (Zamani et al., 2017), perceiving enjoyment, curiosity, adventure, and entertainment (Armstrong et al., 2015). Research has also demonstrated that the social component of clothing exchange events draws clients who are fashion leaders and view exchanging worn apparel as an artistic endeavor (Lang & Armstrong, 2018).

In Italy, swap events are regularly organized in Milan, one of the main fashion cities worldwide, with the project "Swap in the City Milano" founded in 2014 from the great passion of renewing one's wardrobe due to bartering between friends. Maximum 30 individuals participate by booking with an email and being accepted; they pay a limited fee and can bring between one and ten clothing items or accessories in excellent conditions or new (such as dresses, trousers, bags, hats, scarves, and belts). The items will be evaluated by the project staff and split into categories according to their value (cheap, medium, expensive), and for each item/accessory delivered, the participant will receive one token that may be used to "purchase" another item/accessory within the same category of goods (https://swapinthecitymilano.it/).

3.3 Commercial Fashion Sharing Platforms and Subscription Services

In the past, when the exchanges happened in a physical store, whose access was possible through a membership fee, commercial fashion sharing platforms were called "clothing libraries" (Todeschini et al., 2017). Clothing libraries were appreciated by clients for their social aspects by being meeting places for young designers and renters (Pedersen & Netter, 2015; Zamani et al., 2017). Currently, commercial fashion sharing platforms work according to rental or secondhand circular models and, only when platforms are small, they offer networking opportunities and actively engage in decision-making (Cocquyt et al., 2020). In general, however, similar circular business models have a larger size and offer the option as "subscription services" when consumers change the way to use fashion products from a pay-per-product system to an online membership (Camacho-Otero et al., 2019; Tao & Xu, 2018). Through subscription services, customers specify their preferences of price range, size, and kind of desired apparel in an online website, and then, they will receive a tailored box with fashion products and have to decide what to purchase (Tao & Xu, 2018) or return the entire box to the supplier (Day et al., 2020).

Three distinct business models of fashion subscription services have been distinguished: a monthly package of fashion products for a fixed fee; obtaining personal styling services per delivery for a fixed fee; and finally, having access for members to the fashion offerings for home trial without having to make a purchase by paying a one-time fee (Tao & Xu, 2018). Customers sometimes complain about the cancellation process and the lack of social shopping experience; nevertheless, the main benefits recognized of subscription services include the ability to test different styles, excitement, personalization, and cost savings (Camacho-Otero et al., 2019; Day et al., 2020; Vasques et al., 2017). One of the most well-known commercial fashion sharing platforms, the US-based Rent the Runway, founded in 2009 for clothing rental, cooperates with several high fashion brands to provide the selection of four products by paying a rental fee that could be monthly or bimonthly (Liu et al., 2023). Currently, on the Rent the Runway platform is possible also for swapping and buying items so the company's business model involves different circular modes (https://www.renttherunway.com).

3.4 Rental

In the global market, the rental circular business model is growing among companies as proven by the dotted line shown in Fig. 1 indicating the worldwide increase in rental revenue over time (there was only a drop in 2020 due to the COVID-19 pandemic). According to data provider Statista (https://www.statista.com/), the global rental apparel market revenue is expected to reach approximately US$7.45 billion by 2026 (Statista, 2022).

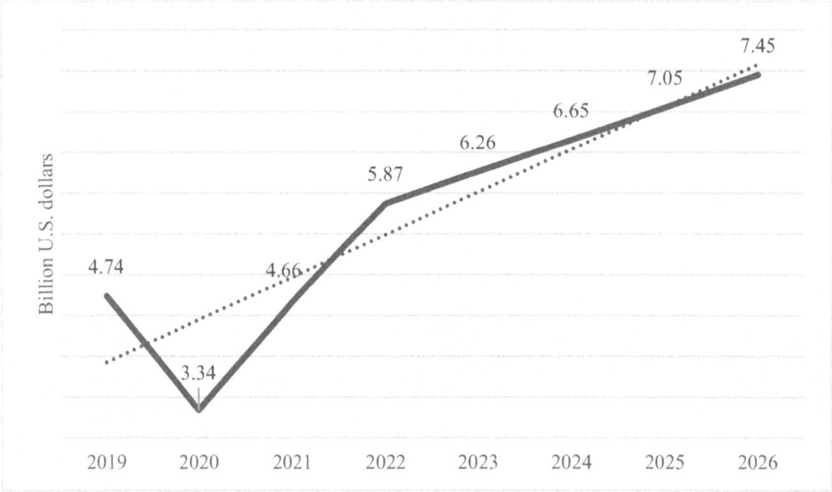

Source: Statista (2022). Revenue of the rental apparel market worldwide from 2019 to 2026. Available at: http://www.statista.com.

Fig. 1 Rental apparel market revenue worldwide (in billion US dollars). (Source: Statista, 2022. Revenue of the rental apparel market worldwide from 2019 to 2026. Available at: http://www.statista.com)

Through rental, a client can have access to a fashion product for a short time by paying a fee and returning it on time and in good condition. Several research papers investigated customer perceptions in terms of adoption and lack of ownership. For some research studies, rental business model enables customers to stay in touch with the dynamic nature of the fashion industry, experiencing new trends for some days at a low and reasonable cost (e.g. Noe & Hyun, 2024; Lang & Armstrong, 2018; Park & Armstrong, 2017; Armstrong et al., 2015). For other studies, the fashion rental success can be associated with sustainability issues and extended life cycle of garments (Becker-Leifhold & Iran, 2018; Lee & Huang, 2020). From the perspective of business opportunity, the rental service can be developed inside an existing fashion company or as a new business activity for an enterprise (Arrigo, 2022). In the first case, the rental provision can generate a market expansion effect for the existing company due to the fact that customers who would not have purchased similar goods for their high price, but with fashion rental they can try them (Feng et al., 2020). In the second case, new entrepreneurs create a rental business model reorganizing their limited resources and entering into partnership with external partners (Adam et al., 2018).

Some global fashion companies launched the rental service in their stores. For instance, H&M was among the first ones to open, in 2019, the Sergels Torg store in Stockholm where clients found the opportunity to rent some beautiful clothes. Currently, the rental service is available in selected H&M stores in the United Kingdom, the Netherlands, Sweden, Germany, and Belgium (H&M Annual Report, 2023). In 2021, Ralph Lauren also announced the launch of "The Lauren Look" platform, as subscription apparel rental initiative with the Lauren Ralph brand, providing consumers an innovative new channel to experience, engage, and ultimately shop the brand (https://corporate.ralphlauren.com).

3.5 Secondhand Resale

The global used apparel market is very interesting since it is expected to grow on average three times quicker than the global apparel market in the coming years (ThredUp Report, 2024). Secondhand apparel business models are often based on platforms that can be immediately created among customers since the possession of used goods is not a prerequisite (Parente et al., 2018). The secondhand resale models are characterized by a permanent transfer of ownership from the seller to the buyer (C2C) or from the seller to a company/platform and then finally to the buyer (B2C) (Arrigo, 2021) and include used goods with, sometimes, other complementary products (Frenken & Schor, 2017; Yrjölä et al., 2021).

Three main groups of motivations drive consumers into buying secondhand fashion: economic motivations, critical motivations, and hedonic and recreational motivations (Machado et al., 2019; Guiot & Roux, 2010; Kim et al., 2021; Abbes et al., 2020). Economic motivations refer to the desire to acquire apparel goods at a lower price for financial reasons, for example, customers have limited economic resources that are not sufficient to purchase new clothing/accessories. In contrast, critical motivations involve consumers' ecological and environmental awareness and a refusal toward a linear system of mass production. Finally, hedonic and recreational motivations can be associated with treasure hunting (the idea of searching for something that is not available to everyone on the market) or social interactions especially in thrift stores which increase satisfaction during the customer experience (Guiot & Roux, 2010) and the desire to feel unique (need to differentiate oneself, to find one's own style and originality or to be accepted within a social group) (Machado et al., 2019).

Online apparel resale is projected to reach US$38 billion by 2027, expanding twice as quickly as secondhand goods overall (ThredUp Report 2023, 2024). In 2023, worldwide, the value of used resale clothing market was projected to US$197 billion, and this value was expected to improve significantly in the upcoming years, rising by almost US$100 billion by 2026, as depicted in Fig. 2 (ThredUp Report 2023, 2024).

A first business example of successful secondhand fashion platform (Abbes et al., 2020) is represented by Vestiaire Collective, founded in France, in 2009, initially as a marketplace C2C for customers to purchase and/or sell pre-owned luxury accessories. However, entrepreneurs quickly understood that for luxury fashion items a C2C platform could be perceived as risky by potential buyers. Therefore, they decided to serve actively as a digital intermediary but also physically verifying the legitimacy and quality of the products. Thanks to this change, Vestiaire Collective became profitable by 2014 (Roussat et al., 2023) and then, in the founded Vestiaire Collective Academy, they trained and educated expert authenticators to recognize the characteristics of genuine materials, methods, and brands—both in the four

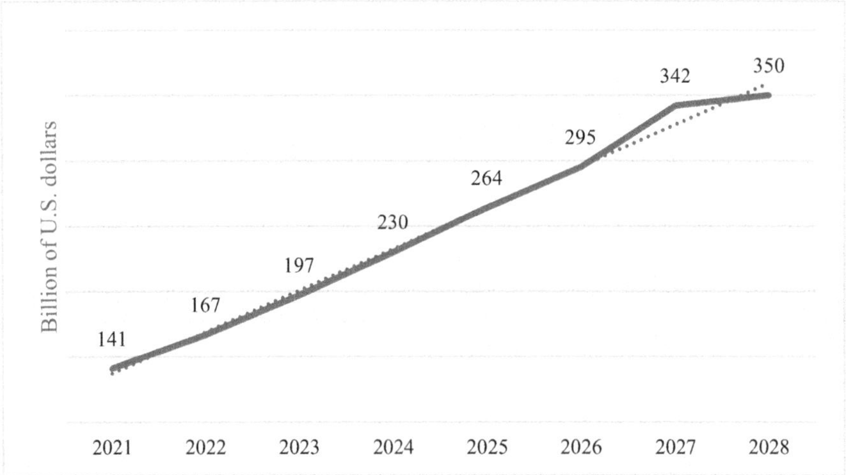

Source: ThredUp (2024), Annual Report, available at: https://www.thredup.com/resale

Fig. 2 Secondhand apparel market value worldwide (in billion US dollars). (Source: ThredUp Report, 2024, Annual Report. Available at: https://www.thredup.com/resale)

authentication centers available in United Kingdom, France, United States, and Hong Kong or online (Vestiaire Collective, Circular Report, 2024).

A second business case of secondhand apparel resale is Zalando, a top online fashion and lifestyle platform in Europe that, founded in 2008 in Berlin, provides customers with multiple propositions from fashion to beauty. Recently, the company decided to offer in 13 markets the sale of pre-owned goods: 'We are pleased to see customers embracing the opportunity to try on and purchase certified second-hand pre-owned articles … in addition to using our dedicated online pre-owned offer' (2023 Annual Report, p. 75). Pre-owned garments go through stringent quality control procedures and are provided to consumers with the same degree of convenience of Zalando, such as free shipping and returns as well as a variety of payment choices. In 2023, a great progress in raising the pre-owned assortment unit's economics and operational efficiencies was reached. This was possible using Zalando's operational synergies such as product data and photos that customers had already bought on their site and incorporating pre-owned into the existent current logistics processes. Zalando offers a selection of about 270,000 items in the pre-owned area with the same level of service as for new products-free delivery and free returns (https://corporate.zalando.com/).

4 Conclusion and Future Research Directions

By reusing, repairing, recycling, or returning resources to maintain them in use over long time, circular business models aim to completely close the apparel product/material cycle and surpass the idea of "end of life" (Henry et al., 2020). Circular business models incorporate the circular economy aspects as guidelines for the design and development of novel business strategies that increase the value to consumers while lowering the overall generated waste and pollution (Abbate et al., 2023). The chapter has provided, from a collaborative consumption perspective, insights into the various circular apparel business models that have been studied in earlier scholarly works (Arrigo, 2021). The described taxonomy of circular apparel business models is relevant to understand the availability of various approaches that apparel managers could try to develop in their circular fashion strategies of product life extension and end-of-life circularity (Dissanayake & Weerasinghe, 2021). In the past years, apparel companies have understood the need to change their business models to become more sustainable with the aim of preserving the natural and social ecosystem. Sustainable and circular business models have started to expand in the apparel industry, and especially secondhand resale, through commercial digital platforms, is expanding quickly. Several business cases have been cited and shortly described to show how circular apparel business models are widespread in the current apparel market.

Further research could better detect the challenges for apparel companies in implementing circular business models in an efficient way and understand the real contribution in terms of ecological sustainability. Additionally, new digital technologies and the spread of artificial intelligence can significantly enhance digital sharing platforms; they could be examined better in the future. Finally, the consumer perceptions of these circular business models could also represent future topics to investigate, for instance analysing the transition from the more historical models toward the current ones. This aspect deserves an in-depth analysis in future academic research.

References

Abbate, S., Centobelli, P., & Cerchione, R. (2023). From fast to slow: An exploratory analysis of circular business models in the Italian apparel industry. *International Journal of Production Economics, 260*, 108824. https://doi.org/10.1016/j.ijpe.2023.108824

Abbes, I., Hallem, Y., & Taga, N. (2020). Second-hand shopping and brand loyalty: The role of online collaborative redistribution platforms. *Journal of Retailing and Consumer Services, 52*(April 2019), 101885. https://doi.org/10.1016/j.jretconser.2019.101885

Abdelmeguid, A., Afy-Shararah, M., & Salonitis, K. (2024). Towards circular fashion: Management strategies promoting circular behavior along the value chain. *Sustainable Production and Consumption, 48*, 143–156. https://doi.org/10.1016/j.spc.2024.05.010

Adam, M., Strähle, J., & Freise, M. (2018). Dynamic capabilities of early-stage firms: Exploring the business of renting fashion. *Journal of Small Business Strategy, 28*(2), 49–67.

Armstrong, C. M., Niinimäki, K., Kujala, S., Karell, E., & Lang, C. (2015). Sustainable product-service systems for clothing: Exploring consumer perceptions of consumption alternatives in Finland. *Journal of Cleaner Production, 97*, 30–39. https://doi.org/10.1016/j.jclepro.2014.01.046

Arrigo, E. (2021). Collaborative consumption in the fashion industry: A systematic literature review and conceptual framework. *Journal of Cleaner Production, 325*, 129261. https://doi.org/10.1016/j.jclepro.2021.129261

Arrigo, E. (2022). Digital platforms in fashion rental: A business model analysis. *Journal of Fashion Marketing and Management, 26*(1), 1–20. https://doi.org/10.1108/JFMM-03-2020-0044

Becker-Leifhold, C., & Iran, S. (2018). Collaborative fashion consumption–drivers, barriers and future pathways. *Journal of Fashion Marketing and Management: An International Journal, 22*(2), 189–208. https://doi.org/10.1108/JFMM-10-2017-0109

Bocken, N. M., Short, S. W., Rana, P., & Evans, S. (2014). A literature and practice review to develop sustainable business model archetypes. *Journal of Cleaner Production, 65*, 42–56. https://doi.org/10.1016/j.jclepro.2013.11.039

Bocken, N. M., De Pauw, I., Bakker, C., & Van Der Grinten, B. (2016). Product design and business model strategies for a circular economy. *Journal of Industrial and Production Engineering, 33*(5), 308–320. https://doi.org/10.1080/21681015.2016.1172124

Camacho-Otero, J., Boks, C., & Pettersen, I. N. (2019). User acceptance and adoption of circular offerings in the fashion sector: Insights from user-generated online reviews. *Journal of Cleaner Production, 231*, 928–939. https://doi.org/10.1016/j.jclepro.2019.05.162

Clube, R. K., & Tennant, M. (2020). Exploring garment rental as a sustainable business model in the fashion industry: Does contamination impact the consumption experience? *Journal of Consumer Behaviour, 19*(4), 359–370. https://doi.org/10.1002/cb.1817

Cocquyt, A., Crucke, S., & Slabbinck, H. (2020). Organizational characteristics explaining participation in sustainable business models in the sharing economy: Evidence from the fashion industry using conjoint analysis. *Business Strategy and the Environment, 29*(6), 2603–2613. https://doi.org/10.1002/bse.2523

Colucci, M., & Vecchi, A. (2024). Tackling climate change with end-of-life circular fashion practices – Remade in Italy with amore. *British Journal of Management, 0*, 1–23. https://doi.org/10.1111/1467-8551.12806

Day, S., Godsell, J., Masi, D., & Zhang, W. (2020). Predicting consumer adoption of branded subscription services: A prospect theory perspective. *Business Strategy and the Environment, 29*(3), 1310–1330. https://doi.org/10.1002/bse.2435

Decoopman, I., Gentina, É., & Fosse-Gomez, M. H. (2010). The generational confusion? Identity issues caused by clothes-sharing between mothers and their teenage daughters. *Recherche et Applications en Marketing (English Edition), 25*(3), 7–27. https://doi.org/10.1177/205157071002500301

Dissanayake, D. G. K., & Weerasinghe, D. (2021). Towards circular economy in fashion: Review of strategies, barriers and enablers. *Circular Economy and Sustainability, 2*, 1–21. https://doi.org/10.1007/s43615-021-00090-5

Feng, Y., Tan, Y. R., Duan, Y., & Bai, Y. (2020). Strategies analysis of luxury fashion rental platform in sharing economy. *Transportation Research Part E: Logistics and Transportation Review, 142*, 102065. https://doi.org/10.1016/j.tre.2020.102065

Frenken, K., & Schor, J. (2017). Putting the sharing economy into perspective. *Environmental Innovation and Societal Transitions, 23*(3), 3–10. https://doi.org/10.1016/j.eist.2017.01.003

Geissdoerfer, M., Pieroni, M. P., Pigosso, D. C., & Soufani, K. (2020). Circular business models: A review. *Journal of Cleaner Production, 277*, 123741. https://doi.org/10.1016/j.jclepro.2020.123741

Gentina, E. (2014). Understanding the effects of adolescent girls' social positions within peer groups on exchange practices. *Journal of Consumer Behaviour, 13*(1), 73–80. https://doi.org/10.1002/cb.1460

Gentina, E., & Fosse-Gomez, M. (2010). Identifying adolescent peer group structure: In the search of social network methods. *Advances in Consumer Research, 9*, 1–24.

Gentina, E., Hogg, M. K., & Sakashita, M. (2017). Identity (re) construction through sharing: A study of mother and teenage daughter dyads in France and Japan. *Journal of Retailing and Consumer Services, 37*, 67–77. https://doi.org/10.1016/j.jretconser.2017.03.008

Guiot, D., & Roux, D. (2010). A second-hand shoppers' motivation scale: Antecedents, consequences, and implications for retailers. *Journal of Retailing, 86*(4), 355–371. https://doi.org/10.1016/j.jretai.2010.08.002

H&M (2023). *Annual and sustainability report*. Retrieved April 2024, from https://hmgroup.com/investors/reports/

Henninger, C. E., Bürklin, N., & Niinimäki, K. (2019). The clothes swapping phenomenon–when consumers become suppliers. *Journal of Fashion Marketing and Management: An International Journal, 23*(3), 327–344. https://doi.org/10.1108/JFMM-04-2018-0057

Henry, M., Bauwens, T., Hekkert, M., & Kirchherr, J. (2020). A typology of circular start-ups: An analysis of 128 circular business models. *Journal of Cleaner Production, 245*, 118528. https://doi.org/10.1016/j.jclepro.2019.118528

Holmes, H. (2018). Material affinities: "Doing" family through the practices of passing on. *Sociology, 53*(1), 174–191. https://doi.org/10.1177/0038038518777696

Iran, S., & Schrader, U. (2017). Collaborative fashion consumption and its environmental effects. *Journal of Fashion Marketing and Management: An International Journal, 21*(4), 468–482. https://doi.org/10.1108/JFMM-09-2016-0086

Jung, S., & Jin, B. (2016). Sustainable development of slow fashion businesses: Customer value approach. *Sustainability, 8*(6), 540. https://doi.org/10.3390/su8060540

Kim, L. N., Woo, H., & Ramkumar, B. (2021). The role of product history in consumer response to online second-hand clothing retail service based on circular fashion. *Journal of Retailing and Consumer Services, 60*(102457), 1–9. https://doi.org/10.1016/j.jretconser.2021.102457

Kumar, L., Kamil, I., Ahmad, M., Naqvi, S. A., Deitch, M. J., Amjad, A. Q., et al. (2022). In-house resource efficiency improvements supplementing the end of pipe treatments in textile SMEs

under a circular economy fashion. *Frontiers in Environmental Science, 10*, 1002319. https://doi.org/10.3389/fenvs.2022.1002319
Lang, C., & Armstrong, C. M. J. (2018). Collaborative consumption: The influence of fashion leadership, need for uniqueness, and materialism on female consumers' adoption of clothing renting and swapping. *Sustainable Production and Consumption, 13*, 37–47. https://doi.org/10.1016/j.spc.2017.11.005
Lang, C., Li, M., & Zhao, L. (2020). Understanding consumers' online fashion renting experiences: A text-mining approach. *Sustainable Production and Consumption, 21*, 132–144. https://doi.org/10.1016/j.spc.2019.12.003
Lee, S. H., & Huang, R. (2020). Exploring the motives for online fashion renting: Insights from social retailing to sustainability. *Sustainability, 12*(18), 7610. https://doi.org/10.3390/su12187610
Liu, N., Lin, J., Guo, S., & Shi, X. (2023). Fashion platform operations in the sharing economy with digital technologies: Recent development and real case studies. *Annals of Operations Research, 329*(1), 1175–1195. https://doi.org/10.1007/s10479-022-04544-3
Machado, M. A. D., Almeida, S. O. D., Bollick, L. C., & Bragagnolo, G. (2019). Second-hand fashion market: Consumer role in circular economy. *Journal of Fashion Marketing and Management: An International Journal, 23*(3), 382–395. https://doi.org/10.1108/JFMM-07-2018-0099
McNeill, L., & Venter, B. (2019). Identity, self-concept and young women's engagement with collaborative, sustainable fashion consumption models. *International Journal of Consumer Studies, 43*(4), 368–378. https://doi.org/10.1111/ijcs.12516
Muthu, S. S. (Ed.). (2024). *Sustainable manufacturing practices in the textiles and fashion sector*. Springer.
Noe, H., & Hyun, J. (2024). Why do and why don't consumers use fashion rental services? A consumption value perspective. *Journal of Fashion Marketing and Management, 28*(3), 566–580. https://doi.org/10.1108/JFMM-07-2023-0168
Parente, R. C., Geleilate, J. M. G., & Rong, K. (2018). The sharing economy globalization phenomenon: A research agenda. *Journal of International Management, 24*(1), 52–64. https://doi.org/10.1016/j.intman.2017.10.001
Park, H., & Armstrong, C. M. J. (2017). Collaborative apparel consumption in the digital sharing economy: An agenda for academic inquiry. *International Journal of Consumer Studies, 41*(5), 465–474. https://doi.org/10.1111/ijcs.12354
Park, H., & Armstrong, C. M. J. (2019). Is money the biggest driver? Uncovering motives for engaging in online collaborative consumption retail models for apparel. *Journal of Retailing and Consumer Services, 51*, 42–50. https://doi.org/10.1016/j.jretconser.2019.05.022
Pedersen, E. R. G., & Netter, S. (2015). Collaborative consumption: Business model opportunities and barriers for fashion libraries. *Journal of Fashion Marketing and Management, 19*(3), 258–273. https://doi.org/10.1108/JFMM-05-2013-0073
Renewcell. (2022). *Annual report*. Retrieved from https://www.renewcell.com/en/renewcell-publishes-its-annual-report-for-2022/
Roussat, C., Carbone, V., & Rouquet, A. (2023). Conceptualizing sharing supply chains–lessons from an exemplary case. *International Journal of Operations & Production Management, 43*(3), 466–488. https://doi.org/10.1108/IJOPM-10-2021-0670
Ruckelshausen, S., & Kamleitner, B. (2024). Hand-me-downs: The wallflower of pre-ownership and a special case for psychological ownership. In P. A. Albinsson, B. Y. Perera, & S. J. Lawson (Eds.), *Understanding collaborative consumption*. Edward Elgar Publishing Ltd. https://doi.org/10.4337/9781035307531.00017
Schor, J. B., & Fitzmaurice, C. J. (2015). Collaborating and connecting: The emergence of the sharing economy. In *Handbook of research on sustainable consumption* (pp. 410–425). Edward Elgar Publishing.
State of Fashion. (2024). Retrieved from https://www.mckinsey.com/industries/retail/our-insights/state-of-fashion

Statista. (2022). Revenue of the rental apparel market worldwide from 2019 to 2026. In *Statista*. Statista Dossier Plus Shared Apparel, Retrieved April 30, 2024, from http://www.statista.com/statistics/1195613/rental-apparel-market-revenue-worldwide/

Statista. (2023). Revenue of the apparel market worldwide from 2018 to 2028 (in trillion U.S. dollars). In *Statista*. DossierPlus Shared Apparel, Retrieved April 30, 2024, from https://www.statista.com/forecasts/821415/value-of-the-global-apparel-market

Tao, Q., & Xu, Y. (2018). Fashion subscription retailing: An exploratory study of consumer perceptions. *Journal of Fashion Marketing and Management: An International Journal, 22*(4), 494–508. https://doi.org/10.1108/JFMM-11-2017-0123

ThredUp Report 2023. (2024). Retrieved May 2024 from https://www.thredup.com/resale

Todeschini, B. V., Cortimiglia, M. N., Callegaro-de-Menezes, D., & Ghezzi, A. (2017). Innovative and sustainable business models in the fashion industry: Entrepreneurial drivers, opportunities, and challenges. *Business Horizons, 60*(6), 759–770. https://doi.org/10.1016/j.bushor.2017.07.003

Tunn, V. S., Bocken, N. M., van den Hende, E. A., & Schoormans, J. P. (2019). Business models for sustainable consumption in the circular economy: An expert study. *Journal of Cleaner Production, 212*, 324–333. https://doi.org/10.1016/j.jclepro.2018.11.290

Vasques, R. A., Koria, M., & Santos, M. C. L. D. (2017). Why do I love you Vaatelainaamo? Analysis of motivations, barriers and opportunities in a Finnish service for sharing clothes. *The Design Journal, 20*(sup1), S721–S731. https://doi.org/10.1080/14606925.2017.1353018

Vestiaire Collective, Circularity Report. (2024). Retrieved April 2024 from https://it.vestiairecollective.com/journal/our-sustainability-manifesto/

Yrjölä, M., Hokkanen, H., & Saarijärvi, H. (2021). A typology of second-hand business models. *Journal of Marketing Management, 37*(7–8), 761–791. https://doi.org/10.1080/0267257X.2021.1880465

Zamani, B., Sandin, G., & Peters, G. M. (2017). Life cycle assessment of clothing libraries: Can collaborative consumption reduce the environmental impact of fast fashion? *Journal of Cleaner Production, 162*, 1368–1375. https://doi.org/10.1016/j.jclepro.2017.06.128

Websites

Decathlon. (n.d.). https://decathlon.com

Hmgroup. (n.d.). https://hmgroup.com/

Kering. (n.d.). https://kering.com

Patagonia. (n.d.). https://patagonia.com

Ralph Lauren Corporate. (n.d.). https://corporate.ralphlauren.com

Renewcell. (n.d.). https://renewcell.com/en

Renttherunway. (n.d.). https://renttherunway.com

Statista. (n.d.). https://statista.com/

Swapinthecitymilano. (n.d.). https://swapinthecitymilano.it/

Vestiairecollective. (n.d.). https://vestiairecollective.com

Zalando. (n.d.). https://corporate.zalando.com/

Circular Economy Dynamics in the Apparel Industry: The Emergence of "Retrovation"

Karolina Crespi Gomes, Amir Maghssudipour, and Silvia Rita Sedita

1 Introduction

The apparel industry, renowned for the heavy environmental impact of the production process, should be at the forefront of a significant transformation, driven by the urgency to meet environmental and social sustainability goals (Pal et al., 2019; Abbate et al., 2023). In order to reinvent the apparel value chain, a variety of actions is expected, not only modifying the upstream stages of the production process, as well as the design, the supplier relations, the product's end of life but also raising the sensitivities of the various stakeholders involved, not underestimating the role played by an increased customers awareness of the side effects of overconsumption, and by the pressures of social movements in rethinking the overall apparel ecosystem toward sustainability (Sedita et al., 2024). The implementation of circular economy practices might be considered one possible solution to accelerate this (Bassett, 2021; Jia et al., 2020). Circular economy, in contrast to the traditional linear economy, entails the adoption of sustainable business models (SBMs) (Bocken et al., 2014) where economic development is separated from resource consumption by reducing the use and recirculating natural resources (Corona et al., 2019). It encompasses designing out waste, maintaining products and materials in use, and regenerating natural systems. However, it is important to recognize that not all firms in the apparel industry are currently implementing circular practices. Despite the widespread awareness of the need for circularity, the adoption of such practices remains uneven, with some companies still adhering to traditional linear models of production.

K. C. Gomes · A. Maghssudipour (✉) · S. R. Sedita
Department of Economics and Management "Marco Fanno", University of Padua, Padua, Italy
e-mail: karolina.crespigomes@unipd.it; amir.maghssudipour@unipd.it; silvia.sedita@unipd.it

S. S. Muthu (ed.), *Circular Business Models in the Apparel Industry*,
Environmental Footprints and Eco-design of Products and Processes,
https://doi.org/10.1007/978-3-031-80193-8_4

The circular economy is grounded in the following four fundamental principles derived from nature (Ellen MacArthur Foundation, 2012; Weetman, 2016; Gardetti, 2019): (1) Waste as nutrients: like in nature, waste becomes food or nutrients. Objects are reused and redesigned to extend their life cycle indefinitely. (2) Resilience: built through diversity, the circular economy uses a wide pool of resources to adapt to different contexts and circumstances. (3) Renewable energy: a system powered by renewable energy ensures a continuous flow of resources, ideas, and information. (4) Systemic approach: interconnected elements create opportunities and set examples for other economies, cultures, and societies.

Building on such principles, we suggest that a notable trend within the shift toward circularity is the resurgence of traditional practices and processes in innovative contexts, a phenomenon which has been called "retrovation" (Suominen & Sivula, 2016). This is particularly relevant in our case because, in addition to innovation, retrovation—by its very nature—frequently involves reviving and modernizing heritage techniques and craftsmanship in ways that can also align with contemporary sustainability objectives (Sedita, 2024). This study explores the complex interplay between circularity and retrovation, aiming to illustrate how this approach can foster sustainable transformations in the apparel industry.

Particularly, by examining the convergence of craftsmanship, traditional production methods, and modern sustainability orientation of firms in the apparel sector, we seek to highlight the potential of heritage practices to drive circular sustainability through innovation. With empirical analysis, we focus on an exemplary case of circular sustainability where we also suggest retrovation is successfully implemented. This case serves as beacons of change, showcasing how integrating traditional knowledge with modern environmental and pro-social goals can lead to significant advancements in sustainability. In fact, our focus on this case is intended to illustrate the broader potential of retrovation to drive change within the apparel sector. While not all firms are currently adopting circular practices, the success story we examine can provide a roadmap for others in the industry to follow.

The focus on this sector is important because it has experienced unprecedented growth, producing between 100 and 150 billion items of clothing annually (McKinsey & Company, 2022; World Economic Forum, 2021). Several businesses in this sector have sped up the season cycles; they manufacture their clothes in other countries, mainly Eastern ones where production costs and labor rights are lower, and they sometimes adopt low levels of transparency in their communications (Gardetti, 2019). This surge in production has doubled the application of textiles since the turn of the millennium, yet garment utilization has drastically decreased, leading to significant waste (MacArthur, 2017). In 2018, the fashion industry was responsible for 2.1 billion tons of CO_2 emissions, roughly 4% of the global total, with some estimates placing this impact as high as 8–10% (Berg et al., 2020). Upstream activities, such as raw material production and processing, contribute around 70% of this impact, while downstream activities like product use account for the remaining 30% (Global Fashion Agenda and McKinsey & Company, 2020).

In response to these environmental and social challenges, some companies in the (global) apparel industry are increasingly adopting eco-fashion practices and sustainable marketing strategies (Joergens, 2006; Fletcher, 2008; Chan & Wong, 2012). This shift is not limited to slow fashion and luxury brands but extends to a part of the fast fashion companies as well, which are known for their cost-efficiency and rapid market turnaround (De Angelis et al., 2017; Kapferer & Michaut-Denizeau, 2014; Torelli et al., 2012). Consumer demand is also a driving force, with a significant segment considering sustainability a key factor in their purchasing decisions (Kotler, 2011).

Thanks to information gathered from primary sources (mainly site visits, participation in seminars, and direct interviews) and secondary sources (mainly sustainability reports and information from company's social media), this chapter focuses on an exemplary case study that exemplifies the concept of retrovation within the apparel industry. This is the case of Rifò,[1] a business in the historical Prato textile district in Italy, where traditional *cenciaioli* artisans have revitalized their practices. These artisans are renowned for their expertise in selecting and repurposing textile scraps and waste, embodying the essence of retrovation by integrating traditional knowledge, innovation, and modern environmental and pro-social objectives.

Through this case study, this research highlights the potential of retrovation to drive circularity-driven sustainability in the apparel industry. It underscores the significance of leveraging traditional knowledge and practices to pave the way toward a more sustainable and circular apparel ecosystem. The revitalization of traditional practices toward innovation, or retrovation, offers a promising pathway for achieving sustainability in an industry often criticized for its environmental and social footprint.

2 Components of Circular Business Models in the Apparel Sector

Circular business models (CBMs) in the apparel industry encompass various components aimed at minimizing waste and maximizing resource efficiency throughout the product life cycle (Dragomir & Dumitru, 2022). CBMs are predicated on the 9R framework of circular economy that stands for the most to the last recommended action being Refuse, Rethink, Reduce, Reuse, Repair, Refurbish, Remanufacture, Repurpose, Recycle, and Recover (Potting et al., 2017). Knowing this, after "refuse" as the main attitude for CBMs, "rethink" is the second in order of importance. In the apparel industry, this means that designing clothes that last longer and are easier to repair are paramount to create a value chain closer to sustainability. The goal is to develop innovative business models that enhance consumer value while maximizing the long-term value of goods, components, and materials. This approach aims to

[1] Rifò is a Tuscan inflection of the verb *rifare* (meaning "to remake, to redo"), a name that takes up the tradition of textile recycling born in Prato and which has been handed down for over a century.

reduce the overall resource consumption, waste, emissions, and pollution (Konietzko et al., 2020).

As articulated by Geissdoerfer et al. (2018), CBMs can be regarded as a subset of sustainable business models (SBMs) due to their sustainability benefits, which stem from reduced resource inputs and lower waste and emissions. This assertion implies that circular supply chain management (CSCM), aimed at promoting sustainable development, should incorporate not only the typical features of a circular economy but also the distinctive characteristics of SBMs. These characteristics include adoption of a long-term perspective, proactive multi-stakeholder management, and creation of additional value through environmental and social sustainability considerations.

Thus, the circular economy represents a departure from the unsustainable, traditional linear fashion model ("take-make-waste"). The infeasibility and risks associated with the linear model are increasingly apparent, given its operational challenges, such as sourcing virgin raw materials, and strategic concerns, like the changing consumer preferences—almost 90% of Europeans believe clothes should be made to last longer (European Commission, 2022). As early as 2018, companies were advised to "self-disrupt their own identity and the sources of their old success" by adopting new business models aligned with sustainability concerns (McKinsey & Company and Business of Fashion, 2022). In this work, we are considering features of CBMs: design for longevity, circular production processes, and material innovation.

2.1 Design for Longevity

One fundamental component of circular business models in the apparel industry involves designing for durability and longevity, ensuring that garments withstand multiple uses and wash cycles (De Falco et al., 2019). Additionally, the utilization of sustainable materials, such as recycled cotton, reduces reliance on virgin resources and minimizes environmental impact (Chua, 2018). However, the recycling of polymeric fabrics is not so sustainable since it has microplastics as sub-products, also generated in the washing process of this clothes (Periyasamy & Tehrani-Bagha, 2022; De Falco et al., 2019; Napper & Thompson, 2016). Knowing this, garments must be designed to last long and for this the choice of the materials is paramount.

Schumann et al. (2021, p. 447) states that "the main problem in textile recycling remains the separation of the different fiber blends", which leads to the reflection that brands must start to design their clothes with mono-material fabrics, but not only this, because there are fabrics that are more sustainable than others based on their durability and composition. Thinking about a broader use of the term redesign, not only the conception of products must have new approaches but also the business models of companies that intends to be sustainable.

2.2 Circular Production Processes

According to the Ellen MacArthur Foundation (2021), a circular economy for fashion fosters the creation of superior products and services for consumers, presenting a significant growth opportunity that not only contributes to a resilient and thriving industry but also regenerates the environment. This regeneration is achieved by reducing waste and pollution, utilizing biodegradable resources, and employing renewable energy. Moreover, it enhances societal well-being (Fig. 1). Essentially, this model emphasizes the reuse of existing resources within the industry: extending the lifespan of clothing; facilitating the transfer of garments between users; and offering products and services that integrate, enhance, and replace garments in line with consumer needs (Ellen MacArthur Foundation, 2021).

However, rethinking the design of fashion supply chains is important to unlocking the potential of these new models. As highlighted by the Ellen MacArthur Foundation (2021), particularly for larger companies, circular business models are often implemented merely as "add-ons" to linear models, whereas they should entail a shift to an entirely new system (Ellen MacArthur Foundation, 2017). Close collaboration with upstream suppliers and retailers is crucial in devising new solutions that increase the longevity of clothing life cycles and retain both control and the additional value of products post-sale (Global Fashion Agenda and McKinsey & Company, 2020).

Nevertheless, textile recycling represents only 1% of the post-consumption treatment. Schumann et al. (2021, p. 184) argued that "[m]echanical recycling of old textiles remains tear fibres are obtained from the remaining old textiles, which are

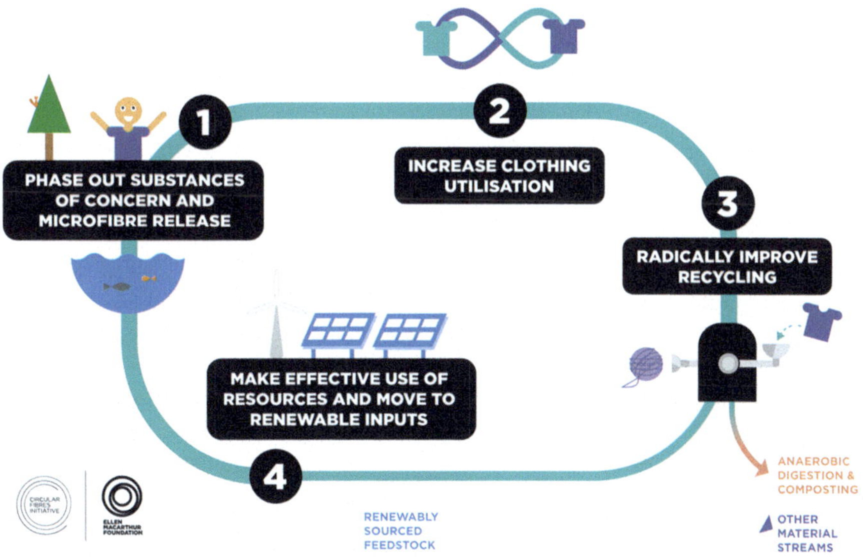

Fig. 1 Circular economy for fashion. (Source: Ellen MacArthur Foundation, 2017)

used almost exclusively in nonwovens for insulation, upholstery and automotive textiles. High-quality applications have so far been niche solutions". More engagement of producers and consumers must exist to enhance the volume of textile recycling and new uses to the recycled pieces must be discovered.

2.3 Material Innovation

In addressing the topic of material innovation for sustainable use in the apparel industry, it is essential to distinguish between natural and manufactured fibers. Natural fibers are derived from plant sources, such as cotton, linen, hemp, jute, and natural bamboo, or animal sources, including wool, silk, and cashmere. Conversely, manufactured fibers are synthesized from raw materials procured from diverse origins. These origins can be natural, as exemplified by fibers such as viscose, modal, and lyocell, or synthetic polymers, such as polyester, nylon, acrylic, polypropylene, and PVC. Contrary to conventional assumptions, natural materials do not inherently represent the superior or more environmentally sustainable option.

Paradoxically, polyester is occasionally marketed as a more sustainable alternative to cotton and wool. This assertion is due to the fact that polyester production necessitates significantly lower water usage compared to cotton—the global average water footprint for producing 1 kg of cotton is approximately 10,000 liters (Freitas et al., 2017), whereas the production of 1 kg of polyester reportedly requires negligible to no water (Fletcher, 2008). Additionally, polyester production does not involve the use of pesticides, further contributing to its perceived environmental benefits.

Nevertheless, the production of manufactured fibers is characterized by an extremely energy-intensive and polluting process. Additionally, the laundering of these materials has been found to be particularly detrimental to the environment, as they shed microplastic fibers that eventually accumulate in seawater (European Commission, 2022). Considering these aspects, numerous fashion brands have begun to introduce "greener" alternatives. These alternatives include organic cotton, which is not only beneficial for the environment but also features a strong social component linked to Fair Trade and ethical production. Furthermore, sustainable synthetics, such as Econyl®—produced through the recycling and purification of ocean and landfill waste have been developed. However, these sustainable options will necessitate significant technological innovations to achieve commercial viability on a large scale.

In this context, it is evident that different regulatory frameworks played a significant role for circularity and the specific case of the apparel sector. The European Green Deal (2019), more than other actions, has been a fundamental milestone. It is a package of policy initiatives, which aims to set the European Union on the path to a green transition, with the ultimate goal of reaching climate neutrality by 2050. It supports the transformation of the EU into a fair and prosperous society with a modern and competitive economy (European Commission, 2019). After this and following the establishment of the Sustainable Development Goals in 2015, some forms of policies for circularity, for businesses, and specifically for the apparel sector have recently followed: important steps have been the Circular Economy Action Plan in 2020; the Ecodesign for Sustainable Products Regulation in 2022; the Sustainability Reporting Directive in 2023; the Green Claims Directive Proposal in 2023; and the Strategy for Sustainable and Circular Textiles in 2023.

3 Collection of Information for the Analysis

The result of two field visits in Prato by one of the author in 2024 is presented, along with two workshops organized by another author at the University of Padova, the first in November 2023, titled "Achieving Carbon Neutrality: Sustainable Business Models, Social Entrepreneurship, and B Corps", with the participation of Rifò's sustainability and innovation specialist, the second in December 2023, within the Sustainable Business Models course, featuring Rifò's founder as guest lecturer.

In addition, we also administered two direct interviews. The first one was conducted with the founder of Rifò in November 2022, when the company was participating in a start-up acceleration program in Florence (promoted by the innovation center Nana Bianca). The second interview was conducted with the firm's sustainability and innovation specialist in May 2024. The time lapse between the two interviews allowed, on the one hand, for the updating of information about the company and, on the other hand, for outlining its development patterns over time.

Finally, we elaborated on two more sources of information. On the one hand, information obtained in an interview in June 2024 with the manager of a textile history museum in Prato, the industrial district where Rifò is located. This interview was specifically important for gaining a broad perspective on the role of *cenciaioli* in the Prato district. On the other hand, we used data provided in the most recent Sustainability Report (2023) by Rifò with a detailed description of members of the company, used materials, obtained certifications, products, business commitment to sustainability and its impact, as well as projects for the future.

4 The Role of Retrovation for Circular Business Models: The Case of Rifò

Rifò is an emerging brand of clothes and accessories committed to producing high-quality products made with recycled textile fibers, all crafted in Italy, specifically in the industrial district of Prato, one of the most iconic mature industrial districts specialized in the textile sector in Europe (Bellandi and Storai, 2022). This company was mentioned by the *Financial Times* among the 1000 fastest-growing companies in Europe, ranking 253rd (Financial Times, 2024). This recognition brings great satisfaction at Rifò but also awareness: in fact, they suggest that economic growth should not be the primary goal. Therefore, in their company's supplementary contract, Rifò has included not only revenue targets but also sustainability and textile recycling goals. As they suggested in Rifò (2024), "[g]rowth and change must go hand in hand".

Utilizing a mechanical and artisanal process inspired by what was developed over a century ago in the textile district of Prato, Rifò transforms fabric leftovers and old clothes into new garments that retain the same quality as the original products. This method has a strong sustainable sense as it enables the recycling of textile waste while significantly reducing the use of water, dyes, and chemicals in production. In a market where it is easy to find affordable and attractive clothing, Rifò believes in doing better by offering high-quality, fashionable attire that also cares for the planet. Their mission is to create a line of stylish and high-quality garments made entirely in Prato from 100% recycled and recyclable fibers.

Established in 2018, the company presently employs 25 people, with 19 being permanent employees. Remarkably, 72% of its workforce is on permanent contracts, and women make up 72% of the staff, with 55% in managerial roles. This gender-inclusive approach highlights Rifò's commitment to a diverse and stable work environment. The company's turnover now exceeds €3 million annually, with 60% of its products sold within Italy and significant exports to Germany, Austria, Switzerland, and France.

Rifò's business model is inherently circular, diverging significantly from the traditional linear models of production and consumption. Key innovations focus on several areas: process innovation through implementing new technologies while maintaining in-house design and outsourcing production; business model innovation by adopting a circular model that revives traditional practices of recycling and reusing materials; and service innovation through developing new customer services that align with sustainable practices. A brief idea of their business model is presented in Fig. 2.

Central to Rifò's strategy is reviving and modernizing heritage techniques and craftsmanship to align with contemporary sustainability objectives. By seamlessly blending tradition with modernity, Rifò integrates time-honored skills like those of the *cenciaiolo* (ragman), shown in Fig. 3, spinner, and weaver into their processes. The *cenciaioli* traditionally used their experience, touch, and manual skills, as well as their sight, to divide bundles of fabrics and fibers into homogeneous groups. In the case under study, they emphasize the reintroduction of traditional mending skills, aiming to create a socially responsible knitting factory. This blend of past and present fosters sustainable innovation, reinforcing the intrinsic link between heritage, sustainability, and innovation.

The retrovated activity of the *cenciaioli* is a specific case of Rifò. In fact, as stated by the manager of a textile history museum in Prato, the activity of the *cenciaioli* has now been lost in Prato and replaced by machinery and automation.

Rifò's commitment to sustainable production is evident across all stages of their operations: products are designed using recycled fibers, with a minimal addition of virgin or organic components to enhance durability. They avoid prints, as garments

Fig. 2 Rifòs's circular business model. (Source: Rifò's website, 2024)

Fig. 3 Rifò's retrovated *cenciaiolo* (ragman). (Source: Rifò's website, 2024)

with prints are less recyclable, and use natural dyes. Additionally, the company minimizes the use of accessories to ensure higher recyclability of their products. The production process involves selecting local artisans who meet Rifò's stringent standards, allowing better quality control and supporting local craftsmanship.

Rifò's supply chain is strictly local, with regular visits to collaborators to ensure adherence to their code of conduct. Bureau veritas conducts semi-announced audits to verify compliance with the International Labour Organization's principles in the field of sustainability. Moreover, the company holds a B-Corp certification since November 2020, underscoring its commitment to high social and environmental performance standards.

In fact, from the interviews, it emerged that Rifò generates 100% of its revenue from goods produced using sustainable methods. Specifically, the company sources raw materials carefully, with sustainability being a key criterion. For instance, they use Global Recycled Standard (GRS) certified materials, which uphold both environmental and social standards. Moreover, it employs innovative natural dyeing techniques, using dyes from peach nuts, walnuts, and chlorophyll pigment, as well as mineral dyes from volcanic rocks. These techniques produce unique, albeit limited, color ranges suitable for small-scale production, adding a distinctive touch to each product. The company also promotes recycling and upcycling. They purchase yarns with recycled fibers and run take-back programs for old jeans and sweaters from customers. Additionally, leftover yarns are used in their *zero waste collection*, creating colorful accessories like hats and gloves.

To engage customers, Rifò uses social media and newsletters to promote and educate about sustainable consumption. In fact, they suggest that their circular service model can further educate and involve customers in sustainable practices. The company also collaborates with other organizations and brands to enhance its sustainable practices, with notable partnerships including Fideuram, Levi's, Bettaknit, and Kering. One outstanding initiative to collaborate with the active participation of consumer in the recycling process is spreading jeans drop-off collection points, as shown in Fig. 4, in purpose-aligned partner stores as NaturaSì and Muji.

Despite its successes, Rifò faces significant challenges, primarily related to scaling up while maintaining sustainability standards. As sustainable fashion grows, maintaining these standards on a larger scale becomes increasingly difficult. Looking ahead, Rifò envisions a significant role for the EU's strategy for

Fig. 4 Rifò's spreading jeans drop-off collection points. (Source: Rifò's website, 2024)

sustainable circular textiles in promoting sustainability. The future of sustainable fashion, according to Rifò, hinges on a shift in consumer habits and business models toward sustainability.

5 Discussion

As introduced in the previous section, we suggest that Rifò exemplifies how traditional practices can be revitalized to foster circularity and sustainable innovation in the modern apparel industry. Through a combination of process, product, and service innovations, the company meets the current demand for sustainable fashion and could set a benchmark for future practices in the industry. By embracing retrovation, Rifò demonstrates the powerful synergy between circularity, innovation, and heritage, showcasing how revisiting and modernizing the interpretation of traditional methods can drive sustainable transformations in the apparel sector.

The transformation toward circularity in the apparel industry, as exemplified by Rifò, demonstrates a shift from traditional sustainable adjustments to innovative business models that integrate heritage techniques with modern circular practices. This shift is significant because it moves beyond mere resource efficiency to encompass a holistic approach that includes manual textile elaboration and natural dyeing processes, which have been largely overlooked in contemporary mass production.

The integration of retrovation within Rifò's business model showcases how traditional practices can be revitalized and adapted to meet modern sustainability goals. This approach not only preserves cultural heritage but also enhances the value proposition of sustainable fashion by creating unique, high-quality products that appeal to environmentally conscious consumers. The emphasis on local artisanship and traditional skills, such as those of the *cenciaioli*, reinforces the importance of community and local economies in achieving sustainability.

Rifò's commitment to sustainability is evidenced in their selection of natural materials such as wool and cashmere, avoiding the generation of microplastics by petroleum-derivative fabrics like polyester and acrylic. Another strategic practice is minimizing the use of accessories to enhance recyclability. With this, Rifò sets a high standard for sustainable fashion. Their supply chain, which is strictly local and adheres to rigorous sustainability criteria, further underscores the feasibility and benefits of a circular economy in the apparel industry.

The empirical evidence from Rifò's case study highlights the practical implications of retrovation in originally driving circularity. The company's ability to blend traditional techniques with innovative business strategies may serve as a benchmark for other firms in the industry. This approach not only addresses the environmental impacts of fashion production but also contributes to social sustainability by supporting local craftsmanship and providing stable employment opportunities.

However, the scalability of such models remains a challenge. As sustainable fashion gains traction, maintaining high standards of sustainability while expanding operations is increasingly complex. Rifò's proactive adaptation of their business model to incorporate sustainability goals alongside economic targets is a crucial strategy in navigating these challenges.

The broader implications of this study suggest that the successful integration of retrovation and circularity can serve as a catalyst for industry-wide change. By demonstrating the viability of sustainable practices rooted in traditional knowledge, Rifò can provide a roadmap for other companies to follow. This case study reinforces the potential of heritage practices to drive innovation and sustainability, offering a promising pathway toward a more circular apparel industry.

Specifically, the discussion on the potential of retrovation within circular business models for the apparel sector reveals several key insights. First, the reintegration of traditional practices not only promotes environmental sustainability but also strengthens cultural heritage and community engagement. The *cenciaioli* artisans in Prato exemplify how historical craftsmanship can be harnessed to meet contemporary sustainability goals.

Moreover, Rifò's approach underscores the importance of local supply chains in achieving sustainability. By sourcing materials locally and maintaining close relationships with artisans, the company reduces its carbon footprint and supports local economies. This localized approach contrasts sharply with the globalized nature of the traditional fashion industry, which often prioritizes cost reduction over environmental and social impacts.

Another crucial aspect of Rifò's strategy is its focus on consumer education and engagement. By involving customers in recycling programs and emphasizing the story behind each product, Rifò fosters a deeper connection between consumers and their purchases. This engagement not only builds brand loyalty but also promotes a broader cultural shift toward more sustainable consumption patterns.

The challenges associated with scaling up such sustainable models cannot be overlooked. As the demand for sustainable fashion grows, companies like Rifò must find ways to expand their operations without compromising their sustainability principles. This requires continuous innovation in production processes, supply chain management, and business strategies. It also necessitates collaboration with other stakeholders, including policymakers, industry peers, and consumers, to create an enabling environment for sustainable growth.

6 Conclusion

In the time of another great challenge driven by hyper-fast fashion and large amounts of textile waste, the apparel industry is looking for new models of circular textile elaboration and production (or even slow fashion). In front of the linear models of extraction, processing, design, manufacturing, retail, use, end of life, and disposal, new phases as introduction of renewable or bio-based inputs, recycling, remanufacturing, reuse, repair, industrial symbiosis, and sharing could be included (Mhatre et al., 2021).

In this context, we introduced the case of retrovation and we analyzed the company Rifò, where this model has been implemented. The case of Rifò demonstrates the transformative potential of combining retrovation with circular economy principles in the apparel industry. By leveraging traditional knowledge and practices, companies can create innovative and sustainable business models that address both environmental and social challenges. More broadly, this company exemplifies the potential of combining traditional practices with modern sustainability initiatives to create a circular economy in the apparel industry.

Rifò's experience provides valuable lessons for the industry, highlighting the need for a holistic approach that integrates sustainability at every level of the value chain. This approach not only addresses environmental and social challenges but also enhances consumer value and supports local economies.

This model highlights the importance of localizing supply chains to minimize carbon footprints and promote local economic strategies. For example, by sourcing materials locally and utilizing traditional artisanal skills, Rifò not only reduces transportation emissions but also supports the livelihoods of local artisans. This approach fosters a symbiotic relationship between sustainability and local culture, enriching the value proposition of sustainable fashion. Furthermore, Rifò's dedication to natural materials and minimizing the use of accessories ensures that their products are not only environmentally friendly but also easier to recycle. This careful consideration of product life cycle from design to disposal exemplifies how circular economy principles can be integrated into every stage of the production process. Such practices set a high standard for the industry and offer a replicable model for other companies seeking to enhance their sustainability credentials.

The broader implications of Rifò's success suggest that the integration of traditional practices and current (and future) sustainability initiatives can serve as a powerful catalyst for industry-wide change. By demonstrating that heritage techniques can interact with innovative business strategies, Rifò can provide a roadmap for other companies to follow.

However, the scalability of such models remains a significant challenge. As sustainable fashion gains traction, maintaining high standards of sustainability while expanding operations becomes increasingly complex. Rifò's ability until now to navigate these challenges through continuous innovation and adaptation can serve as a crucial strategy for other firms. It is relevant that these companies engage in ongoing research and development to refine their processes and maintain their commitment to sustainability.

Consumer education and engagement are also critical components of this transition. Efforts to involve customers in recycling programs and emphasize the story behind each product help foster a deeper connection between consumers and their purchases. This engagement not only builds brand loyalty but also promotes a broader cultural shift toward more sustainable consumption patterns. Educated consumers are more likely to support sustainable brands, creating a virtuous cycle that benefits both the industry and the environment.

The pursuit of sustainability in the apparel industry is not an isolated endeavor. All stakeholders in the production and consumption chain must share responsibility.

Despite the current limited presence of very stringent directives and regulations from governments and international organizations for mass production, and the low level of motivation and awareness among the majority of consumers to curb consumerism, some pioneering companies committed to sustainability must proactively adapt their business models. These adaptations aim to reduce negative externalities on the environment and society.

Relevant implications emerge. Particularly, policymakers must create an enabling environment through supportive regulations and incentives, while industry peers should collaborate to share best practices and drive collective progress. Consumers, too, play a pivotal role by making informed purchasing decisions and supporting brands that prioritize sustainability.

This study contributes to the growing body of evidence supporting the viability and benefits of circular business models, offering a roadmap for others in the industry to follow in the pursuit of a more sustainable future. The collective effort of all stakeholders, businesses, consumers, and policymakers will be crucial in driving this transformation and ensuring a sustainable and prosperous future for the apparel industry. From the theoretical point of view, this work offers one of the rare empirical studies on the concept of retrovation, which has recently emerged as an important pattern in the more established field of responsible innovation for addressing grand societal challenges (Sedita, 2024). Specifically, it provides a specific case study of retrovation in the realm of circularity and sustainability. Moreover, based on an innovative interpretation of practices typical of certain local heritages, it also suggests a preliminary link between the concept of retrovation and studies of local economic development where productive traditions and territorial/sectorial identities play a fundamental role.

There are limitations in this study. As previously mentioned, the case of Rifò suggests tools for applying retrovation by mixing traditional techniques and practices with innovation from a sustainability perspective. However, it cannot be claimed that other companies or the entire sector operate in the same manner. Therefore, the case under analysis might serve as an example for other companies in the apparel sector for the future. This also suggests interesting avenues for future research, which include expanding the sample of interviewed companies—also in other contexts—and empirically validate the assertions made in this chapter, thereby enhancing the practical applicability of our theoretical framework.

Acknowledgments This work was supported by the University of Padova, Department of Economics and Management, Project SID 2023 "Resolving the Retrovation Paradox: Leveraging the Past for a Sustainable Future—PAST4FUT".

References

Abbate, S., Centobelli, P., & Cerchione, R. (2023). From fast to slow: An exploratory analysis of circular business models in the Italian apparel industry. *International Journal of Production Economics, 260*, 108824.

Bassett, N. (2021). Sustainable fashion through circular business innovations: New business models reduce waste. In A. Matthes, K. Beyer, H. Cebulla, M. G. Arnold, & A. Schumann (Eds.), *Sustainable textile and fashion value chains*. Springer. https://doi.org/10.1007/978-3-030-22018-1_16

Bellandi, M., & Storai, D. (2022). Structural change and agency in territorial development: The case of mature industrial districts. *Regional Studies*, 1–13.

Berg, A., Granskog, A., Lee, L., & Magnus, K. H. (2020). *How the fashion industry can urgently act to reduce its greenhouse-gas emissions*. McKinsey & Company Report.

Bocken, N., Short, S. W., Rana, P., & Evans, S. (2014). A literature and practice review to develop sustainable business model archetypes. *Journal of Cleaner Production, 65*, 42–56. https://doi.org/10.1016/j.jclepro.2013.11.039

Chan, T., & Wong, C. W. Y. (2012). The consumption side of sustainable fashion supply chain: Understanding fashion consumer eco-fashion consumption decision. *Journal of Fashion Marketing and Management: An International Journal, 16*(2), 193–215.

Chua, J. M. (2018). The future of recycled cotton denim leans on innovation. *Sourcing Journal Media*. https://sourcingjournal.com/denim/denim-innovations/recycled-cotton-deniminnovation-109974/

Corona, B., Shen, L., Reike, D., Carreón, J. R., & Worrell, E. (2019). Towards sustainable development through the circular economy – A review and critical assessment on current circularity metrics. *Resources, Conservation and Recycling, 151*, 104498.

De Angelis, M., Adigüzel, F., & Amatulli, C. (2017). The role of design similarity in consumers' evaluation of new green products: An investigation of luxury fashion brands. *Journal of Cleaner Production, 141*, 1515–1527.

De Falco, F., Di Pace, E., Cocca, M., et al. (2019). The contribution of washing processes of synthetic clothes to microplastic pollution. *Scientific Reports, 9*, 6633. https://doi.org/10.1038/s41598-019-43023-x

Dragomir, V. D., & Dumitru, M. (2022). Practical solutions for circular business models in the fashion industry. *Cleaner Logistics and Supply Chain, 4*, 100040. https://doi.org/10.1016/j.clscn.2022.100040

Ellen MacArthur Foundation. (2012). *Towards the circular economy*. Ellen MacArthur Foundation.

Ellen MacArthur Foundation. (2017). A New Textiles Economy: Redesigning fashion's future. http://www.ellenmacarthurfoundation.org/publications/

Ellen MacArthur Foundation. (2021). *Circular Business Models: Redefining growth for a thriving fashion industry*. https://ellenmacarthurfoundation.org/fashion-business-models/overview

European Commission. (2019). *The European green deal*. https://commission.europa.eu/strategy-and-policy/priorities-2019-2024/european-green-deal_en

European Commission. (2022). *EU Strategy for sustainable and circular textiles*. https://eur-lex.europa.eu/resource.html?uri=cellar:9d2e47d1-b0f3-11ec-83e1-01aa75ed71a1.0001.02/DOC_1&format=PDF

Financial Times. (2024). Available at: https://www.ft.com/ft1000-2024

Fletcher, K. (2008). *Sustainable fashion and textile products*. Sustainable Fashion and Textiles.

Freitas, A., Zhang, G., & Mathews, R. (2017). Water Footprint Assessment of polyester and viscose and comparison to cotton. *Water Footprint Network* (pp. 1–89). https://waterfootprint.org/media/downloads/WFA_Polyester_and__Viscose_2017.pdf

Gardetti, M. A. (2019). Introduction and the concept of circular economy. In S. S. Muthu (Ed.), *Circular economy in textiles and apparel* (pp. 1–11). Woodhead Publishing.

Geissdoerfer, M., et al. (2018). Business models and supply chains for the circular economy. *Journal of Cleaner Production, 190*, 712–721. https://doi.org/10.1016/j.jclepro.2018.04.159

Global Fashion Agenda and McKinsey & Company (2020). Fashion on climate. How the fashion industry can urgently act to reduce its greenhouse gas emissions.

Jia, F., Yin, S., Chen, L., & Chen, X. (2020). The circular economy in the textile and apparel industry: A systematic literature review. *Journal of Cleaner Production, 259*, 120728.

Joergens, C. (2006). Ethical fashion: Myth or future trend? *Journal of Fashion Marketing and Management, 10*(3), 360–371.

Kapferer, J. N., & Michaut-Denizeau, A. (2014). Is luxury compatible with sustainability? Luxury consumers' viewpoint. *Journal of Brand Management, 21*(1), 1–22. https://doi.org/10.1057/bm.2013.19

Konietzko, J., Baldassarre, B., Brown, P., Bocken, N., & Hultink, E. J. (2020). Circular business model experimentation: Demystifying assumptions. *Journal of Clean Production, 277*, 122596. https://doi.org/10.1016/j.jclepro.2020.122596

Kotler, P. (2011). Reinventing marketing to manage the environmental imperative. *Journal of Marketing, 75*(4), 132–135.

MacArthur, E. (2017). Beyond plastic waste. *Science, 358*(6365), 843.

McKinsey & Company and Business of Fashion. (2022). *The state of fashion 2023*. Available at: https://www.mckinsey.com/industries/retail/our-insights/state-of-fashion

Mhatre, P., Panchal, R., Singh, A., & Bibyan, S. (2021). A systematic literature review on the circular economy initiatives in the European Union. *Sustainable Production and Consumption, 26*, 187–202.

Napper, I. E., & Thompson, R. C. (2016). Release of synthetic microplastic plastic fibres from domestic washing machines: Effects of fabric type and washing conditions. *Marine Pollution Bulletin, 112*(1–2), 39–45. https://doi.org/10.1016/j.marpolbul.2016.09.025

Pal, R., Shen, B., & Sandberg, E. (2019). Circular fashion supply chain management: Exploring impediments and prescribing future research agenda. *Journal of Fashion Marketing and Management: An International Journal, 23*(3), 298–307.

Periyasamy, A. P., & Tehrani-Bagha, A. (2022). A review on microplastic emission from textile materials and its reduction techniques. *Polymer Degradation and Stability, 199*, 109901. https://doi.org/10.1016/j.polymdegradstab.2022.109901

Potting, J., Hekkert, M. P., Worrell, E., & Hanemaaijer, A. (2017). Circular economy: Measuring innovation in the product chain. *Planbureau voor de Leefomgeving, 2544*.

Rifò. (2024). Available at: https://www.linkedin.com/posts/rifolab_la-sfida-pi%C3%B9-grande-%C3%A8-crescere-economicamente-activity-7179097996196802560-aZYE?utm_source=share&utm_medium=member_desktop

Schumann, A., et al. (2021). Sustainability and digitalization in the global textile value chain. In A. Matthes et al. (Eds.), *Sustainable textile and fashion value chains*. Springer Nature. https://doi.org/10.1007/978-3-030-22018-1_23

Sedita, S. R. (2024). Responsible innovation for addressing grand societal challenges: The role of social innovation, exaptation, and retrovation. *European Planning Studies, 32*, 1–23.

Sedita, S.R., Crespi Gomes, K., Pellegrin, V. (2024). The evolution of the academic discourse on sustainability in the fashion industry: insights from a bibliometric analysis. *Consumption and Society* (published online ahead of print 2024). Retrieved Aug 28, 2024, from https://doi.org/10.1332/27528499Y2024D000000023.

Suominen, J., & Sivula, A. (2016). Retrovation–the concept of a historical innovation. *WiderScreen, 19*(3–4).

Torelli, C. J., Özsomer, A., Carvalho, S. W., Keh, H. T., & Maehle, N. (2012). Brand concepts as representations of human values: Do cultural congruity and compatibility between values matter? *Journal of Marketing, 76*(4), 92–108.

Weetman, C. (2016). *A circular economy handbook for business and supply chains: Repair, remake, redesign, rethink*. Kogan Page Publishers.

World Economic Forum and Boston Consulting Group. (2021). *Net-zero challenge: The supply chain opportunity*.

Karolina Crespi Gomes holds a PhD in Business Administration from the University of Vale do Itajaí (Brazil), taught the course "Sustainable Marketing and Innovation," and served as a teaching assistant for the course "Sustainable Business Models" at the University of Padova (Italy). She has been interested in zero waste consumption behavior and circular economy practices.

Amir Maghssudipour is an assistant professor of management at the Department of Economics and Management, University of Padova. He obtained a Ph.D. in development economics and local systems in 2019. His research interests include issues at the intersection of sustainability and innovation within firms’ networks and territorial systems like industrial districts and clusters. Recently, his research has delved into the realm of renewable energy communities, exploring their role in driving sustainability transitions.

Professor Silvia Rita Sedita obtained a Ph.D. in economics and management of firms and local systems in 2004. She is a full professor of management at the University of Padova. Her current research agenda includes issues within the field of management of sustainable innovation in inter-organizational networks.

Modeling Social Circular Economy Business Models in the Fashion Industry: The Atelier Riforma Case Study

Francesco Marengo, Dario Cottafava, Elena Ferrero, and Laura Corazza

1 Introduction

The circular economy (CE) is an economic system designed to reduce waste and optimize resource use by maintaining materials in continuous circulation. Unlike the traditional linear economy, which follows a "take-make-dispose" model, CE seeks to create closed-loop systems where products and materials are reused, repaired, and recycled, thereby preserving their value throughout their life cycle (Ghisellini et al., 2016; Kirchherr et al., 2017). Circular economy business models (CEBMs) operationalize these principles, focusing on strategies that cycle, extend, intensify, and dematerialize material and energy loops to reduce resource inputs and waste emissions (Moreno et al., 2016; Ünal et al., 2019). These models are designed not only to generate economic value but also to address environmental sustainability by promoting the efficient use of resources (Daou et al., 2020).

Despite the extensive literature on CE and its environmental benefits, there is a notable research gap concerning the social implications of CE and how circular and

F. Marengo · L. Corazza
Department of Management "Valter Cantino", University of Turin, Turin, Italy
e-mail: francesco.marengo@unito.it; laura.corazza@unito.it

D. Cottafava (✉)
Faculty of Economics and Business, Business Department, Universitat Autònoma de Barcelona, Cerdanyola del Vallès, Barcelona, Spain

Centre d'Estudis i de Recerca en Emprenedoria i Innovació Social (CREIS), Universitat Autònoma de Barcelona, Bellaterra, Cerdanyola del Vallès, Barcelona, Spain
e-mail: dario.cottafava@uab.cat

E. Ferrero
Atelier Riforma s.l.r, Turin, Italy
e-mail: elena@atelier-riforma.it

S. S. Muthu (ed.), *Circular Business Models in the Apparel Industry*, Environmental Footprints and Eco-design of Products and Processes,
https://doi.org/10.1007/978-3-031-80193-8_5

social dimensions interact within business models (García-Muiña et al., 2021; Pitkänen et al., 2023). While some studies highlight job creation as a social benefit of CE, a comprehensive analysis of how circular practices can be integrated with social missions within business models remains underexplored. Addressing this gap is crucial, especially in industries with significant environmental and social impacts, such as the textile and fashion industry.

The textile and fashion industry is characterized by its resource-intensive nature and substantial environmental footprint, driven by the fast fashion model that promotes rapid production and consumption cycles (Todeschini et al., 2017). This industry faces multiple environmental challenges, including high energy and water usage, pollution from dyes, and significant textile waste (Peters et al., 2021; Saha et al., 2021). Additionally, social issues such as poor working conditions, insufficient wages, child labor, and forced labor are prevalent in the global supply chains of this sector (Desore & Narula, 2018; Köksal et al., 2017). At the end of its value chain, the traditional linear model encourages a cultural mass consumption intersected with inequalities and stratification (Warde, 2015). Social circular business models (SCBMs) in the textile sector have the potential to simultaneously address these environmental and social challenges by integrating sustainable practices with social missions aimed at improving labor conditions and promoting social equity (Arribas-Ibar et al., 2022; Beyer & Arnold, 2022).

This chapter aims to theorize the emergent social circular economy (SCE) by analyzing a case study focused on Atelier Riforma, an innovative Italian start-up in the apparel sector. Atelier Riforma integrates circular economy principles with a strong social mission, seeking to mitigate the environmental and social impacts of the fashion industry. By employing semi-structured interviews, direct observations, and internal documents, this study provides insights into how Atelier Riforma operationalizes CE principles intertwined with a social mission. The analysis focuses on material, financial, and impact flows to demonstrate how SCBMs can generate financial value while simultaneously creating social and environmental benefits.

The structure of this chapter is as follows: the next section provides a comprehensive literature review on CE, SCBMs, and their impact assessment methods. Following this, the methodology section outlines the single-case study approach adopted for this research, including data collection and analysis procedures. The results section presents the findings from the case study, highlighting the integration of social and environmental dimensions in Atelier Riforma's business model. The discussion section critically examines these results in light of existing literature, addressing the challenges and implications of implementing SCBMs in the textile sector. Finally, the conclusion discusses the limitations of this study and suggests directions for future research, emphasizing the need for further conceptualization and empirical analysis of SCBMs across different sectors.

By investigating the case of Atelier Riforma, this chapter contributes to the understanding of how SCBMs can address dual objectives of environmental sustainability and social equity, offering valuable insights for researchers, practitioners, and policymakers interested in fostering sustainable development (SD) through innovative business models.

2 Literature Review

2.1 Circular Economy

The circular economy (CE) is an economic system designed to minimize waste and optimize resource use by keeping materials in circulation for as long as possible. It aims to reduce the extraction of natural resources and waste by preserving the value of products and materials throughout their life cycle (Calisto Friant et al., 2020). To achieve this, the CE model seeks to create a closed-loop system in which materials are reused, repaired, and recycled rather than discarded after their initial use. In particular, Suárez-Eiroa et al. (2019) found that the literature proposes three main strategies within the CE paradigm: (i) minimizing inputs of raw materials and outputs of waste, (ii) keeping resource value as long as possible within the system, and (iii) reintegrating products into the system when they reach the end of their life. This approach contrasts sharply with the traditional linear economy based on the "take-make-waste" approach, a neoclassical model that focuses on efficient resource allocation without considering the finite nature of natural resources, resulting in resource depletion and environmental degradation derived from waste disposal and pollution, which could limit the ability of future generations to satisfy their needs (Ghisellini et al., 2016).

Considering that the CE framework offers an alternative to the linear economy model, authors have argued that the CE transition is emerging as one of the most popular societal responses to environmental problems and to achieving more sustainable development worldwide (Kevin van Langen et al., 2021). For this reason, Kirchherr et al. (2017) argue that the CE can be viewed as an approach to implement the much-discussed concept of sustainable development, defined as development that meets the needs of the present without compromising the ability of future generations to meet their own needs (Brundtland & Khalid, 1987). Sustainable development is often framed in the triple bottom-line approach, which highlights that sustainable development should simultaneously promote economic development, preserve the environment, and tackle societal issues concerning human beings and their needs (Elkington, 1998). However, the interconnections and integration of sustainable development and CE are highly discussed in the literature. On the one hand, Schroeder et al. (2019) suggest that the circular economy can directly help attain a variety of the United Nations (UN) Sustainable Development Goals (SDGs), while for Korhonen et al. (2018), a successful circular economy contributes to all three dimensions of sustainable development: economic, environmental, and social. Both circular economy and sustainable development rely on multidisciplinary or interdisciplinary approaches to better integrate noneconomic aspects into development and emphasize intra- and intergenerational commitments and cooperation among stakeholders (Geissdoerfer et al., 2017). On the other hand, some authors also highlight contradictions and gaps in the integration of the two concepts. Geissdoerfer et al. (2017) highlight that sustainability aims to equally benefit the environment, economy, and society, treating these three spheres horizontally. In contrast, CE hierarchically prioritizes economic systems, with primary benefits for economic actors and secondary environmental benefits like reduced resource depletion and pollution, while social benefits are often indirect, such as increased manual labor and fairer taxation (Calisto Friant et al., 2020).

Despite differing views on the integration and interrelation of CE and sustainable development (SD), an advantage of CE is that it can be implemented and analyzed at various levels. While SD, considered by some as a vague concept and difficult to implement by economic actors (Naudé, 2011), primarily focuses on a global perspective, CE can be implemented at the micro-, meso-, and macro-levels (Kirchherr et al., 2017). At the microlevel, CE involves single firms or organizations and even specific product-level changes and consumer preferences. The meso-level focuses on regional implementations involving multiple organizations, such as eco-industrial parks, where industrial symbiosis is the most studied form of CE implementation (Merli et al., 2018). Finally, the macro-level addresses national and global policies and industry structures. An effective transition toward the CE paradigm requires changing traditional patterns across all three levels, from policymakers and regulators to individual firms and even individuals (Dey et al., 2020). Scholars have predominantly studied CE at the macro-level, followed by the microlevel and meso-level (Merli et al., 2018).

At the microlevel, CE involves measuring reduced ecological footprints, direct financial value through material recovery, and growth through new business models (Lewandowski, 2016). The transition of companies toward CE, although in its early stages, requires significant changes in business and managerial practices, including efficient use of energy, materials, and resources to reduce environmental impact (Ünal et al., 2019). This has led to the conceptualization of circular business models (CBMs), which are crucial for the circular economy to flourish in the long run (Geissdoerfer et al., 2020). CBMs are defined as models that cycle, extend, intensify, and/or dematerialize material and energy loops to reduce resource inputs and waste emissions within an organizational system (Geissdoerfer et al., 2020). The value proposition in CBMs is grounded in maintaining the economic value embedded in products after their use and repurposing it for new market offerings, which requires a set of return flows from end users to producers, often through intermediaries (Rosa et al., 2019). The value proposition in CBMs is grounded in maintaining the economic value embedded in products after their use and exploiting it for new market offerings, which requires a set of return flows from end users to producers, often through intermediaries (Selvefors et al., 2024). Based on value proposition typology, Rosa et al. (2019) identify five common CBM archetypes: "share," which focuses on maximizing goods usage and reducing waste; "regenerate," aimed at enhancing the earth's bio-capacity; "optimize," which improves process efficiency; "loop," which involves recycling practices; and "virtualize," which includes solutions that dematerialize systems or deliver them virtually. Besides the type of value proposition, determinants of CBM creation are determined by endogenous and exogenous factors. Assmann et al. (2023) identify eight determinants for introducing CBMs: culture, regulation, market, strategy, business case, collaboration, operations, and knowledge. To understand the specific characteristics of CBMs, tools such as the EcoCanvas, a practical tool for rethinking, developing, and integrating environmental exchanges into business models help entrepreneurs with a visual tool that explicitly integrates circular economy strategies into different layers of the model (Daou et al., 2020).

2.2 *Social Circular Economy*

Given the relevance that CE may play in promoting sustainable development, and the uncertainties regarding how CE can promote social equity or measure its social benefits (Millar et al., 2019), scholars have recently begun to address the social dimension of the circular economy. Social equity and social aspects are crucial for sustainable development (Souza Piao et al., 2023); hence, for CE to fully promote sustainable development, investigating its social implications is essential. However, considering that only very recently scholars have begun addressing this topic (Pitkänen et al., 2023), aspects regarding the social dimension of sustainable development in CE are still notably underdeveloped, under-conceptualized, and under-researched, particularly in terms of social inclusion (Souza Piao et al., 2023). According to literature reviews by Padilla-Rivera et al. (2021) and Mies and Gold (2021), most studies focusing on the social aspects of the CE do not cover a wide range of perspectives or attempt to build a holistic framework but rather have a narrower focus on impacts such as employment or job creation, education, health and safety, and democratic participation. Both studies emphasize the need for broader inclusion of social aspects in the CE discourse.

From a conceptual point of view, Clube and Tennant (2020) investigated the embedded consideration of human needs within seminal texts on circular economy-related concepts, such as the Blue Economy, Cradle-to-Cradle (C2C), Regenerative Design for Sustainable Development, and the Performance Economy. They found that these concepts' original ideas incorporated certain satisfiers of human needs, which seem to have become lost in CE practical forms of implementation. For this reason, Kirchherr (2021) has proposed a conceptual framework to implement circular economy strategies considering the human side through the concept of "circular justice." At the macro-level, most studies focus on public health and safety, job creation, social justice, and sociocultural capital (Mies & Gold, 2021). However, societal issues are mainly approached as corollaries for the circular economy, such as ensuring societal inclusion of informal workers in CE entities (Guibrunet et al., 2017) or the role of active societal participation combined with clear political goals in driving societal transformation toward a circular economy.

At the meso-level, studies often focus on supply chains. For example, Masi et al. (2017) performed a systematic review of 77 documents discussing the meso-level of supply chains in relation to the circular economy. However, most of these studies marginally address social aspects (Souza Piao et al., 2023). Vahidzadeh et al. (2021) conducted a literature review on regional industrial symbiosis, identifying six thematic aspects, one of which dealt with social aspects, specifically highlighting collaborations among stakeholders of symbiosis networks as a key component for successful industrial symbiosis. Additionally, Vanhuyse et al. (2021) found that the current discourse on circular cities is limited, primarily focusing on governance issues related to legal frameworks and citizen participation in waste management systems, with social impacts largely unaddressed.

At the microlevel, Walker et al. (2021) studied CE industry perspectives on the social dimension and its assessment through an analysis of multiple CE companies in Italy and the Netherlands. They found that although the social dimension was considered important, the actual assessment of social impacts was limited. Regarding business models, besides sustainable business models built on the triple bottom-line framework, no study has yet been developed on the conceptualization of holistic social CBMs, but there are a few studies addressing their social implications (Salvador et al., 2020) or analyzing the links between CBMs and social business models (Fehrer & Wieland, 2021). The existing literature on CBMs addresses the importance of social networks around which these models are built (Susur & Engwall, 2023) or advocates for the social benefits of CBMs by influencing the behavior of customers and end users toward upcycling, recycling, and reusing (Lüdeke-Freund et al., 2024). An important body of literature on the social implications of CBMs concerns the assessment and measurement of their societal impacts. Multiple disciplines have assessed this issue and developed specific methods, which will be explored in the following section. Overall, many studies on CE focus on practical stages represented by tools and practical measures (Mies & Gold, 2021). However, critical discussions on the circular economy's theoretical foundations involving the integration of and contribution to the social dimension as intended by sustainable development have yet to take place, as they are often studied peripherally (Souza Piao et al., 2023).

2.3 *Impact Assessment of Social and Circular Business Model*

Schwandt (2015) highlights that evaluating and assessing business models is inherently complex, requiring continuous engagement with current evaluation theory and practice. This complexity is amplified for social and CBMs. While traditional, linear business models primarily focus on economic performance (Kanzari et al., 2022), circular economy models integrate diverse approaches and disciplines aimed at creating regenerative systems (Hidalgo-Carvajal et al., 2021). In particular, regarding metrics for the evaluation of CBMs, Saidani et al. (2024) identified current challenges such as the actual usage and adoption of indicators and methods for measuring CE, the comparison of different products and companies, and the integration of material scarcity, renewable materials, and quality tools. They also emphasize that future studies should address linking social indicators to the evaluation of CE. Evaluation tools for these models must capture success across multiple dimensions, particularly when business models address comprehensive social, environmental, and economic challenges, as seen in sustainability-oriented business models (Mattos et al., 2022). As a result, various tools and methods have been developed to assess the impact of circular and social business models separately.

For evaluating the social and circular value, various methods assess environmental sustainability, one of which is the Cradle-to-Cradle (C2C) certification. The C2C certification, a nongovernmental ecolabel issued by the "McDonough Braungart Design Chemistry" firm, is a multi-attribute, continuous improvement methodology, evaluating products across five categories of human and environmental health (Llorach-Massana et al., 2015). Built on the C2C concept, it focuses on three principles (Braungart et al., 2007). The first principle, "waste equals food," designs systems where waste is a nutrient for nature or other industrial processes. The second principle promotes using energy from "current solar income," emphasizing renewable energy. The third principle, "celebrate diversity," integrates cultural, economic, and environmental aspects. To achieve the certification, companies must meet requirements across five categories: material health, product circularity, clean air and climate protection, water and soil stewardship, and social fairness (C2C Institute, 2021). These categories are scored on a five-grade scale: basic, bronze, silver, gold, and platinum, reflecting increasing stringency in requirements. For example, the C2C certification demands the use of 100% electricity coming from renewable energy to achieve the highest grade (platinum) in the "clean air and climate protection" category, among other requirements (C2C Institute, 2021).

The Cradle-to-Cradle (C2C) certification, while valuable, has been critiqued for not holistically addressing environmental aspects of products from a full life-cycle perspective (Llorach-Massana et al., 2015). It primarily assesses environmental impacts at a micro (product) level, with only one out of five categories evaluating social fairness (Niero et al., 2018). In contrast, B Corps offer a case that move from product-level to company-level assessments Incorporating both social and environmental impacts in their assessment, B Corps represent a new form of prosocial enterprising that integrates the purpose of serving the common good into the legal framework of the business (Muñoz et al., 2018). They are businesses that meet the highest verified standards of social and environmental performance, public transparency, and legal accountability, balancing profit and purpose. To be certified as a B Corp, a company must be assessed by B Lab through the B Impact Assessment (BIA) and achieve a score greater than 80 out of 200 points, evaluating five areas: workers, community, environment, customers, and governance (Nigri & Del Baldo, 2018). Becoming a B Corp is not merely an ethical decision but a strategic one, with significant implications for various stakeholders. The BIA and compulsory reporting activities can attract investors and appeal to consumers who prefer certified socially responsible companies, as well as influence entrepreneurial funding decisions (Clark & Babson, 2012). Despite the existence of various reporting standards like the Global Reporting Initiative (GRI), small and medium-sized enterprises (SMEs) have struggled to adopt them due to the time and resources required. The BIA is simpler and less time-intensive than traditional reporting standards like the GRI, as it does not require a materiality analysis or extensive stakeholder engagement during evaluation and prioritization (Shields & Shelleman, 2017). Nonetheless, BIAs have proven to be effective in assessing social and environmental impacts, as

Nigri and Del Baldo (2018) found that the social and environmental commitments of B Corps are directly integrated into company management, and the BIA results support top management decision-making and strategic planning.

A final evaluation method that comprehensively supports the assessment of societal and environmental impacts is the theory of change (ToC) (Mayne, 2017). ToC is primarily used to identify and assess positive impacts, explaining how and why impact happens by mapping pathways that connect interventions to results and providing guidelines on measuring realized impacts (Mayne, 2015). To do so, usually a flowchart based on a logic model that connects activities to short-term outputs, outcomes, and finally long-term impacts is developed to operationalize the theory of change and to understand how to select indicators for measuring the impacts. A key benefit of ToC is its utility in both planning and evaluation phases, illustrating the resources used (inputs) to implement strategies designed to achieve desired results, which helps in identifying impacts through logically connected steps (outputs, outcomes, and impacts) (Margoluis et al., 2009). Originating from social change traditions (Connell & Kubisch, 1998), ToC's malleability has allowed its application in assessing both social and environmental impacts of various different projects and interventions (Anderson, 2018). For instance, Dale et al. (2023) conducted a ToC-based evaluation of a transport intervention in the United Kingdom, finding it effective for assessing social, economic, and environmental impacts of public infrastructural interventions. Michelini et al. (2020) applied ToC to a food-sharing platform business model, identifying and assessing positive impacts of the food sharing platform such as environmental benefits derived from food waste reduction and social benefits derived from feeding impoverished people with unsold food. Mattos et al. (2022) applied ToC to a circular business model case in the metal industry, identifying and measuring economic benefits through cost modeling of recycled versus nonrecycled components, environmental benefits from reduced CO_2 emissions, and social benefits from higher-quality jobs compared to previously adopted linear models.

Given their increased adoption (Rossi & Luque-Vílchez, 2021), also driven by new legal requirements (Abela, 2022), it is important to consider the role of sustainability accounting and reporting practices in assessing and disclosing the social and environmental dimensions (Scarpellini et al., 2020). Numerous standards for sustainability accounting are available for social, circular, and linear businesses (Scarpellini et al., 2020). The "GRI Standards," which have become the leading guideline for voluntary reporting (Halkos & Nomikos, 2021), are built on the triple bottom-line framework, requiring disclosure of economic, social (e.g., safety, diversity), and environmental (e.g., waste, emissions) aspects. These standards are widely adopted by CBMs (Opferkuch et al., 2021). However, most sustainability reporting standards do not fully integrate the circular economy. For instance, CE is discussed in the recently revised "GRI 306: Waste 2020," effective for reports published on or after January 1, 2022. Scarpellini (2022) analyzed how businesses report social impacts related to CE in their sustainability reporting and stakeholder relationships, finding that most professionals operating in CE view CE business models as having positive societal impacts. These impacts include developing new professional opportunities, improving public health due to increased environmental quality and less pollution, and enhancing environmental education for young people. However, the author also argues that most companies adopting CE principles do not report specific societal impacts achieved through closing material loops in a circular model. Additionally, accounting researchers and professional bodies focus on environmental sustainability and reporting but have given limited attention to the concept of circularity (Aureli et al., 2023). These practices are becoming increasingly important as they are also becoming legally required by legislation.

Though this brief presentation of assessment techniques does not aim to be exhaustive, it illustrates the potential tools available to comprehensively address the environmental and social dimensions of circular and social business models.

2.4 CE in the Textile Sector

The fashion industry has experienced significant growth over the past decades, particularly with the rise of fast fashion, which emphasizes rapid acquisition and disposal of mass-produced, standardized fashion items (Fletcher, 2010). This model promotes widespread consumption of easily replaceable garments but has substantial environmental and social drawbacks (Todeschini et al., 2017). The clothing and footwear industries are resource-intensive, with approximately 75% of the energy demand occurring before the garments reach retail (Peters et al., 2021). Many of these products are produced in low-wage countries using nonrenewable energy, leading to high environmental impacts. Additionally, the industry is water-intensive, as it requires cotton grown in arid areas requiring extensive irrigation (Thorisdottir & Johannsdottir, 2020) and causes significant water pollution from untreated dye discharge (Saha et al., 2021). From an end-of-life perspective, the European Union generated an estimated 6.95 million tons of textile waste in 2020, with a significant portion ending up in landfills or incinerators. Only 4.4 kg per person were collected separately for reuse and recycling (Deckers et al., 2024). Despite pressure on the fashion industry to address environmental and social issues, low-cost production methods remain prevalent, exacerbating consumerism (Thorisdottir & Johannsdottir, 2020).

Considering these multiple environmental challenges, researchers have identified the CE as a potential solution to mitigate the negative impacts of the textile industry (Abdelmeguid et al., 2024). Saha et al. (2021) analyzed the relationship between five fields of CE actions (take, make, distribute, use, and recover) and the three dimensions of sustainability performance in the textile sector, finding that raw material sourcing and manufacturing still require significant improvements to become environmentally friendly. Abdelmeguid et al. (2024) conducted a comprehensive analysis of management practices, identifying eight key themes essential for driving circular and sustainable practices in the fashion industry: collaborations, marketing strategies, product-related characteristics, education and awareness, business models, integration of technology, regulations and legislation, and strategic planning. In particular, they found that collaboration with suppliers, recyclers, waste management organizations, and consumers is crucial for enhancing efficiency, circularity, and innovation across the value chain. Waste management is also critical, a field in which Solis et al. (2024) found that "preparing for reuse" is the most beneficial managerial practice for waste reduction from both environmental and socioeconomic perspectives. Therefore, it is not surprising that the second-hand market has been highlighted as a best practice of sustainable responsibility for businesses and consumers (D'Adamo et al., 2022).

While much of the CE research in the textile sector focuses on environmental aspects, the industry also faces significant social challenges (Thorisdottir & Johannsdottir, 2020). Worker conditions are a major concern, especially in low-wage countries, where fashion brands outsource much of their production (Desore & Narula, 2018). Other social risks include child labor and extensive working hours (Köksal et al., 2017). This has led to increased research and regulatory developments in controlling the textile and clothing industry supply chain (Silva et al., 2023). Corporate social responsibility (CSR) practices in this sector are often driven by legal requirements (Nayak et al., 2022), even though they are usually limited in national boundaries even if the supply chain is globalized. Some studies on CBMs in the textile industry also consider their social implications. For example, Barletta et al. (2024) found that leasing practices in the textile industry can address social issues by ensuring ethical conditions for workers and reducing environmental impacts through lower raw material usage and overproduction. Another example is the study by the Impact Institute (2019), which quantifies the "true price" of a pair of jeans by including external social and environmental costs. They estimate that social externalities like insufficient wages, health and safety issues, and child labor, along with environmental externalities like pollution and climate change, would add approximately $30 to the retail price.

Despite studies in the literature arguing for the potential for circular innovations to improve financial performance (Dainelli et al., 2024), there is a gap in the literature regarding how CBMs can positively impact society beyond mitigating the negative societal issues of traditional clothing and textile supply chains by addressing job conditions (Bubicz et al., 2021). This highlights the need for further research to explore the comprehensive social benefits of implementing circular economy practices in the textile industry.

3 Materials and Methods

3.1 Single-Case Study Research

In this chapter, we adopt a single-case study methodology. A case study is an empirical inquiry that investigates a contemporary phenomenon within its real-life context, especially when the boundaries between the phenomenon and context are not clearly evident (Yin, 2018). Unlike other research strategies, such as experiments that isolate the phenomenon from the context to analyze specific variables, case studies are a form of research that is deliberately contextually informed. This approach is often used in business, managerial, and accounting studies because it allows for an in-depth analysis of businesses and various organizations in their natural settings, helping to understand the nature and complexity of the object of study (Voss et al., 2002). Case studies enable exploratory analyses that address the growing frequency and magnitude of changes in technology, business practices, models, and methods. The rich detail provided by case studies increases the likelihood of generating novel theories and creative insights (Cooper & Morgan, 2008).

Our study is based on a single-case study, where the authors sought to describe the fundamental elements characterizing a phenomenon and the relationships that are established among them (Puntillo et al., 2021). According to Yin (2018), single case offers significant insights to contribute to knowledge and theory building by confirming, challenging, or extending the theory. The single-case study approach is particularly appropriate as it allows us to delve deeply into the complexities and nuances of the phenomenon under investigation within its real-life context, providing rich insights and a comprehensive understanding that other methodologies may not offer.

In our case, the transition toward a circular economy in the fashion and textile sector is still in its early stages, and the comprehensive environmental and societal impacts of organizations adopting a socio-circular business model have yet to be fully explored. Therefore, we employ an in-depth case study approach to gain a comprehensive understanding of this phenomenon. The case study methodology is preferable for exploring new phenomena and allowing for serendipitous findings in idiosyncratic situations (Eisenhardt & Graebner, 2007). An in-depth case study aligns with the goal of this research because studying a single firm in depth enables the development of intimate familiarity, which enhances our ability to extract insights from the data. Additionally, the complex and multidimensional nature of the analysis requires a comprehensive investigation, which is more feasible within a single stand-alone entity (Dyer & Wilkins, 1991). Case selection is based on a theoretical convenience sampling strategy (Voss et al., 2002). Following this strategy, we selected a start-up conceived with a circular business model in the textile and fashion sector operating in Turin, Italy. Founded in 2020, this case stands out as an early adopter of a business model that facilitates circular practices within the textile and fashion industry.

We developed the case study based on multiple qualitative data sources, primarily interviews and archival documents. Using various information sources allows for triangulation, enhancing the reliability of the findings (Bowen, 2009). Document analysis included materials from the company's online website, internal documents, a business model canvas, and a social and environmental impact report developed by the start-up. Semi-structured interviews were conducted with the CEO and her collaborator of the start-up, as well as with the CEO of a second circular firm operating downstream in the value chain of the case. This second interview allowed us to increase data sources on our case and thus to avoid post-hoc rationalizations and increase the reliability of our study (Ünal et al., 2019). Semi-structured interviews, chosen for their flexibility and ability to elicit crucial insights (Qu & Dumay, 2011), lasted approximately 1 h and were conducted online. No materials or questions were shared prior to the interview, though transcripts were provided to the interviewees afterward. Questions focused on the circular value proposition, social contributions, opportunities and weaknesses of the business models, barriers and drivers to scaling up socio-CBMs in the textile sector, and the identification and evaluation of the start-up's environmental and social impacts, particularly their interaction. Table 1 shows the interview questions. To ensure the methodological rigor, we have developed a research protocol according to Yin (2018), shown in Table 2.

3.2 *Case Study Description*

Atelier Riforma was conceived between 2018 and 2019 when the two cofounders met during a postgraduate educational program on entrepreneurship aimed at recent graduates interested in launching innovative start-ups. Sharing similar visions and ideas, they developed the concept during the course, leading to the formal establishment of Atelier Riforma as an innovative social start-up in 2020. The start-up's mission is to mitigate the environmental and social impact of the fashion industry through a circular economy approach.

Initially, Atelier Riforma operated on a business-to-consumer (B2C) model, focusing on collecting, cataloging, and upcycling of used garments. Individuals and organizations donated discarded clothing, which was then carefully cataloged, recording all its characteristics (condition, composition, type, person/entity of origin, etc.). After this process, garments were divided into those suitable for the secondhand market and those needing to be fixed and repaired before being ready to be sold (upcycling). The latter were distributed to a network of over 25 tailors,

Table 1 Interview questions

When and how was the start-up conceived?
What is the vision and mission of the start-up?
What is the business model and has it changed over time?
What are the strengths, weaknesses, opportunities, and threats of your business model?
What are the main barriers for the scale-up of the circular start-up?
What is the social value proposition of the start-up?
To what extent are the social and environmental dimensions interrelated in the start-up operations and impacts?
Who are the beneficiaries of your positive social and environmental impacts?
How do you assess and evaluate social and environmental impacts?
How and to what extent are you able to measure the integration of social and environmental impacts?
What are the barriers to social and environmental impact assessment and evaluation?

Table 2 Research protocol

Planning and design	*Research question:* How are the social and environmental impacts interrelated in a start-up with a social circular business model in the textile sector? Methodological approach: Single-case study *Case study selection:* The case was selected based on a theoretical convenience sampling strategy (Voss et al., 2002), given the relevance of the case for the research question and accessibility to data collection
Data collection	Data collection was developed in three phases. The first phase involves a critical review of the literature on the topic of social CBMs and their assessment and evaluation, with a focus on the textile sector. This phase is necessary to analyze the data in the context of existing literature and is presented in Sect. 2. The second phase concerns the analysis of the start-up's public documents and archives (including its website) and internal strategic documents, including a business model canvas, an environmental and social impact assessment based on the theory of change, and a feasibility study of the start-up's operations. The third phase includes the interview with the start-up's CEO and her collaborator, along with a second interview with the CEO of a second circular firm operating downstream in the value chain of the case actor to better triangulate findings and gather an external informed perspective on the case
Data analysis	The interviews were transcribed and analyzed concurrently by the authors, who developed a coding system to capture the main themes and insights resulting from the interviews
Presentation of the results	The results were presented in light of the literature reviewed in the first phase of data collection. Through this analysis, the identified main characteristics of the start-up, which qualify it as a start-up adopting a social circular business model, are presented in the results section through a business model canvas, the results of theory of change analysis, and a material-impact-financial flow analysis developed by the authors

designers, and social tailoring workshops across Italy, including social enterprises that promoted labor inclusion of disadvantaged individuals such as migrants, ex-prisoners, and women victims of violence. The upcycled garments were sold on an e-commerce platform, with Atelier Riforma retaining a percentage of the sales. Despite its innovative approach, this model proved financially unsustainable due to the labor-intensive process, high marketing costs, and challenges in attracting investors. Recognizing these limitations, the founders innovated their business model by designing Re4Circular in 2021 (subsequently succeeding in fully developing the solution during the years 2022 and 2023). Re4Circular is an AI-based technology designed to automate the classification and sorting of used garments, addressing the inefficiencies of the initial business model. By photographing each garment and its label, the AI extracts essential data such as color, type, season, and material composition. This information determines the most suitable circular destination for each garment, whether for reuse, upcycling, recycling, or others. Once classified, garments are uploaded on a digital platform, connecting entities and companies with pre- or post-consumer clothing, such as collection and sorting organizations, with circular fashion businesses that need these materials for resale, recycling, or upcycling activities.

Re4Circular operates as a business-to-business (B2B) digital marketplace, facilitating transactions between garment collectors and circular fashion businesses. The classification technology linked to the platform ensures transparency and efficiency by providing detailed images and information about each garment, allowing businesses to make informed purchasing decisions. This system eliminates the traditional bulk buying of garments sight unseen, which often leads to inefficiencies and additional sorting post-purchase. Instead, circular businesses can search for and purchase only the garments that meet their specific needs, significantly improving the overall efficiency of the circular fashion value chain.

Atelier Riforma's vision is to transform the fashion industry into a sustainable model based on circular economy principles. Their mission is to minimize the use of virgin resources and reduce textile waste by promoting the reuse, upcycling, and recycling of post-consumer garments. By leveraging Re4Circular, Atelier Riforma aims to create a transparent, efficient, and scalable system that maximizes the value of used textiles and supports the fashion industry‘s transition to a circular economy. Currently, Atelier Riforma operates from Turin, Italy, with a team that includes a chief executive officer, a chief operating officer, a chief marketing officer and a collaboration with two tech companies: one for AI development and the other for the platform and web app development. For financial sustainability, the platform charges a fee on sales and plans to introduce an annual subscription fee for the users once the classification technology is fully developed. By providing an automatic classification technology and a digital marketplace for the wholesale commerce of pre- and post-consumer unused textiles, Atelier Riforma contributes to enhancing the overall sustainability of the fashion industry. In addition to their core operations, Atelier Riforma develops a service to monitor textile waste collection in urban areas, aimed at municipalities. This service will provide transparency regarding the quantities and destinations of recovered textiles, further supporting sustainable practices at a municipal public level.

Since its conception, Atelier Riforma has transitioned from a labor-intensive B2C model to a scalable B2B model powered by AI technology, facilitating efficient and transparent transactions in the circular fashion value chain. This innovative approach not only supports environmental sustainability but also promotes social equity by involving disadvantaged individuals in system of stakeholders engaged. The start-up continues to evolve, aiming to drive systemic change in the fashion industry toward a more circular and sustainable model. The start-up's long-term vision includes promoting a fully regenerative fashion industry where garments are designed to last and be easily repaired, reused, transformed, or recycled. By ensuring that no garment becomes waste, Atelier Riforma aims to create economic value, generate new employment opportunities, and significantly reduce the environmental impact of the fashion industry.

4 Results

To investigate how the social and environmental impacts are interrelated in its social circular business model, different assessment methods are used and presented in this results section. First, a social circular business model adapted from Robinson (2017) is used to understand the status social, environmental, and economic value proposition, mapping its key business functions. Then, an impact assessment based on ToC is presented in Sect. 4.2 and finally, an impact-material-financial flow analysis is presented. The results are critically discussed in the discussion section (Sect. 5) in light of the relevant literature.

4.1 Social Circular Business Model

The social circular business model canvas of Atelier Riforma (Fig. 1) was developed based on a traditional business model framework independently created by the company along with the development of the technology Re4Circular and the consequent introduction of the new B2B business model. It was then adapted to the social circular business model canvas developed by Robinson (2017), highlighting not only the economic value proposition but also the social and environmental aspects. This

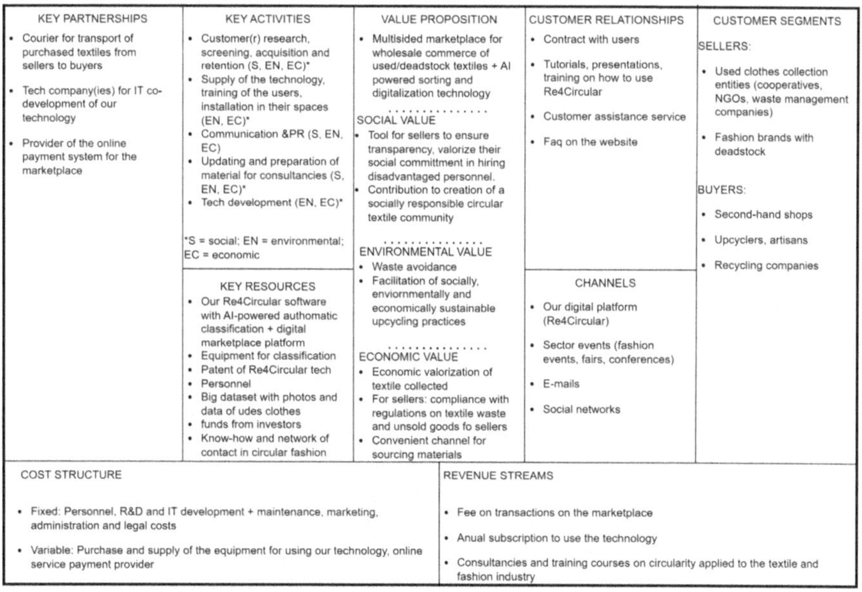

Source: Authors' elaboration inspired by Robinson (2017)

Fig. 1 Atelier Riforma social circular business model canvas. (Source: Authors' elaboration inspired by Robinson, 2017)

adaptation involved mapping key activities according to their contributions across social, environmental, and economic dimensions.

To understand the three dimensions of Atelier Riforma's value proposition, it is necessary to analyze its key activities, which have been mapped in the social circular business model canvas based on their social, environmental, and economic contributions. The environmental implications are deeply embedded in their business model, as their technology is designed to facilitate transactions among circular entities in the textile sector. Specifically, Atelier Riforma promotes waste avoidance by enabling efficient and economically viable transactions among businesses that recycle, upcycle, or reuse textiles. This system enhances transparency and efficiency in the supply chain for circular garments. For this reason, the environmental dimension covers all the key activities: customer research and retention to engage the actors of the textile and fashion industry in circular practices; the supply of the technology (and related activities such as tech development) to facilitate circular transactions, communication, and public relations to increase awareness on the topic; and the preparation of materials for consultancies on the circular economy.

The social dimension is particularly evident in multiple aspects. First, it emerges in their efforts to engage clients within the value chain who are committed to social responsibility. These clients ensure good working conditions for disadvantaged groups, and Atelier Riforma's technology supports these businesses, promoting social inclusion in the sector. Therefore, the activities related to customer research, screening, acquisition, and retention become activities with intrinsic social value as they facilitate business for socially responsible organizations promoting social inclusion through hiring disadvantaged people in good working conditions. Second, Atelier Riforma's continuous public information and dissemination activities on responsible consumption and the circular economy in fashion help promote greater awareness among citizens and positively influence their behavior (inform, buy less, choose secondhand, use longer, repair, swap, donate, discard responsibly, etc.). Finally, Atelier Riforma provides its technology free of charge to nonprofit organizations that collect used clothes for solidarity purposes, allowing them to use it for clothes unsuitable for these users. This leads to two positive results: the organization will not have to bear the cost of disposing of the unused garments and will instead earn from their resale to circular businesses, thereby having more income to support its solidarity activities. Considering these social impacts, Atelier Riforma also contributes to the development of a socially responsible circular community involving key organizations in the value chain, contributing to the development of a socially responsible branch of the textile and fashion industry.

Regarding economic value, Atelier Riforma aims at capturing a part of the economic value it generates through the facilitation of transparency, automation of sorting activities, and valorization of the intrinsic value of garments and textiles which, if treated and commercialized properly, can become a new resource instead of waste. Atelier Riforma captures this economic value through transaction fees on the marketplace and subscription fees to use the technology they provide. This core business activities are also complemented by consultancies and training courses on circularity applied to the fashion industry.

In general, the main contribution to the reduction of environmental impacts in the fashion and textile sector is constituted by waste avoidance and potentially reduced exploitation of natural resources, as the broader the reach of the circular textile business becomes, the less textile waste is generated and the fewer resources are depleted for new textile production (assuming that average quantity consumption does not variate in the transition from linear to circular goods). Additionally, there is a societal impact resulting from the creation of a socially responsible community among value chain actors. This community's work and social responsibilities are recognized and valorized, reinforcing the integration of social and environmental values in the business model.

4.2 Theory of Change Impact Assessment

The theory of change analysis conducted by Atelier Riforma is a prospective analysis, meaning that the actual impacts have not yet been measured. This initial planning outlines the expected impacts that Atelier Riforma aims to achieve and how they will be measured. The primary aim is to transform the fashion industry's production model from a linear, wasteful system to a sustainable, circular economy model. Atelier Riforma plays multiple roles: connector, resource provider, scaler, and activist. These roles synergistically contribute to changing the system by addressing the inefficiencies in managing post-consumer textiles.

Atelier Riforma's theory of change outlines their vision of a future where post-consumer textiles are no longer waste but valuable materials for recycling, upcycling, and reuse. The ultimate outcomes they seek include the valorization of post-consumer textiles in the circular economy and the establishment of a system where the specific characteristics of textiles are identified and shared. Intermediate outcomes necessary to achieve this vision include widespread consumer demand for sustainable and circular fashion and legislative support through regulations and incentives for circular practices. In the short term, Atelier Riforma aims to focus on activism to raise awareness and drive demand for circular fashion. They plan to enhance their technology to speed up the cataloging process, making it more attractive for collectors and selectors of used textiles. This includes improving their marketing strategy to reach more circular buyers and facilitate the efficient and transparent management of post-consumer clothing. The direct impacts expected in the short term include increased efficiency in sorting and cataloging used garments, enhanced transparency in the supply chain, and greater economic viability for circular transactions. Through its activities, Atelier Riforma aims to impact various categories of stakeholders, including entities managing textile waste, circular fashion businesses, municipalities, and citizens. These impacts are anticipated to lead to long-term outcomes such as reduced textile waste, decreased exploitation of virgin resources, and the creation of a socially responsible community within the fashion industry. Figure 2 shows a visual flowchart summarizing the logic model behind Atelier Riforma's theory of change.

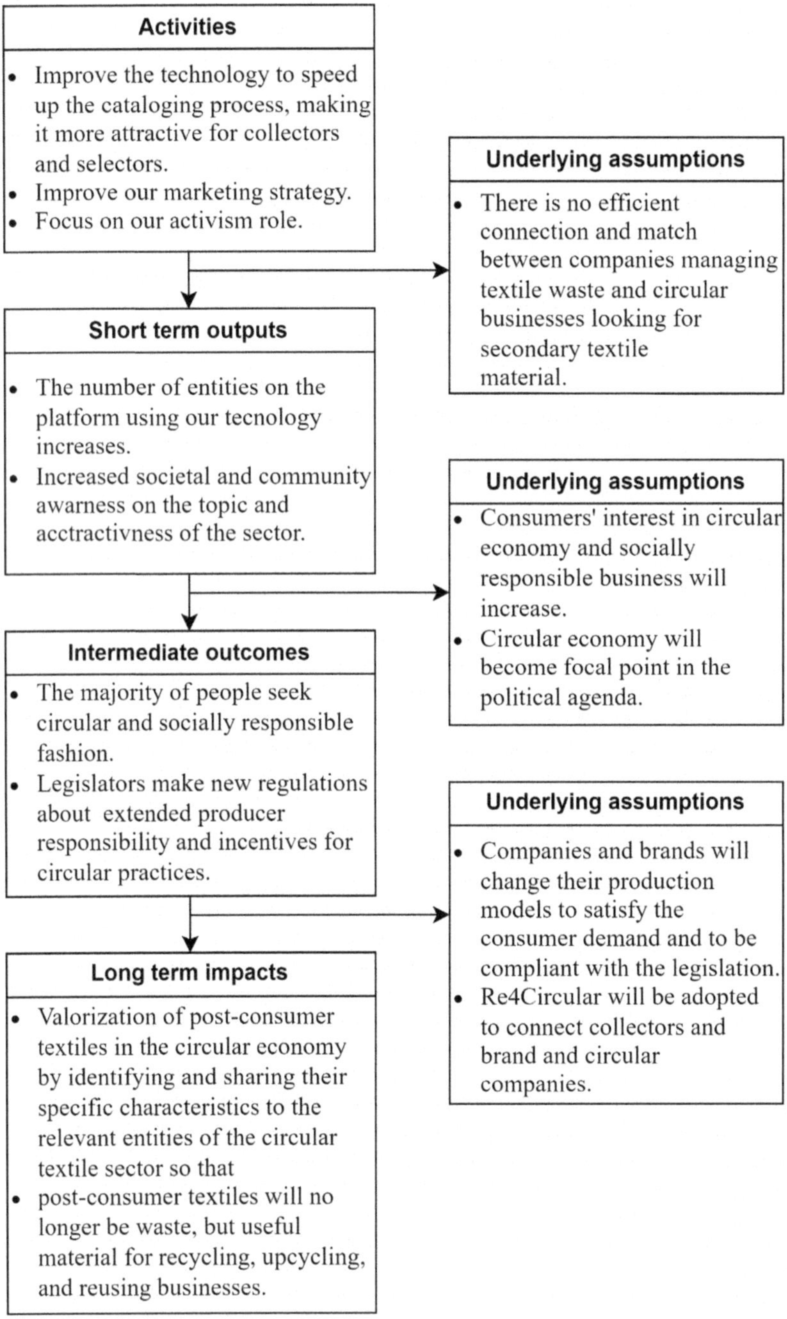

Source: authors' elaboration

Fig. 2 Flowchart of Atelier Riforma's theory of change. (Source: Authors' elaboration)

To measure these impacts, Atelier Riforma has identified several key indicators. These indicators will help determine whether the expected impacts have been achieved. For environmental impacts, they will track the number of used clothes and deadstock fabrics cataloged and uploaded on the platform, the materials they are made of, and the proportion of clothes sold versus those available on the marketplace. Additionally, they will monitor the impact on transport pollution by assessing the stimulation of the local market and the reduction of global transport distances resulting in reduced CO_2 emission. For socioeconomic impacts, they will assess how the platform affects local circular businesses, the increase in the market for recycled/upcycled/reused clothes, the promotion of artisanal competencies that fast fashion has been eroding, and the number of citizens engaged through social media platforms through the dissemination and activist activities. They will also track the number of new local employees hired by circular companies and organizations due to Re4Circular.

Atelier Riforma's approach involves continuously refining their technology and strategies based on these indicators to ensure they are on the right track toward achieving their long-term vision. The success of their theory of change relies on their ability to validate the assumptions that consumer demand and legislative support will drive the adoption of circular practices and that their technology will effectively connect collectors and circular businesses.

4.3 Financial Material Impact Flow Analysis

To understand the flows of environmental, social, and economic impacts activated and generated by Atelier Riforma, we conducted a material and impact flow analysis. This analysis maps material (environmental benefits mainly from waste avoidance), impact (positive social impact), and financial (economic value) flows. We elaborated on two flows: one based on Atelier Riforma's initial B2C business model and a second flowchart focusing on the B2B business model based on Re4Circular technology.

In this analysis, depicted in Fig. 3, the three diverse flows are mapped across the main steps of Atelier Riforma's value chain. While it does not constitute a fully developed material/substance flow analysis (Huang et al., 2012), the material flows represented in the figure show the environmental benefits derived from waste and resource depletion avoidance, which run across all stages of both business models. The flowchart also highlights how social impacts are internally generated by Atelier Riforma in the B2C business model, whereas in the B2B model, they are indirectly promoted through the technology and valorization of socially responsible partners.

The flowchart further maps the financial flows. In the B2C business model, financial value is directly captured by the start-up from the final consumers and backwardly runs across the steps of the value chain. On the other hand, in the B2B model, the financial value is indirectly captured from final stakeholders. The

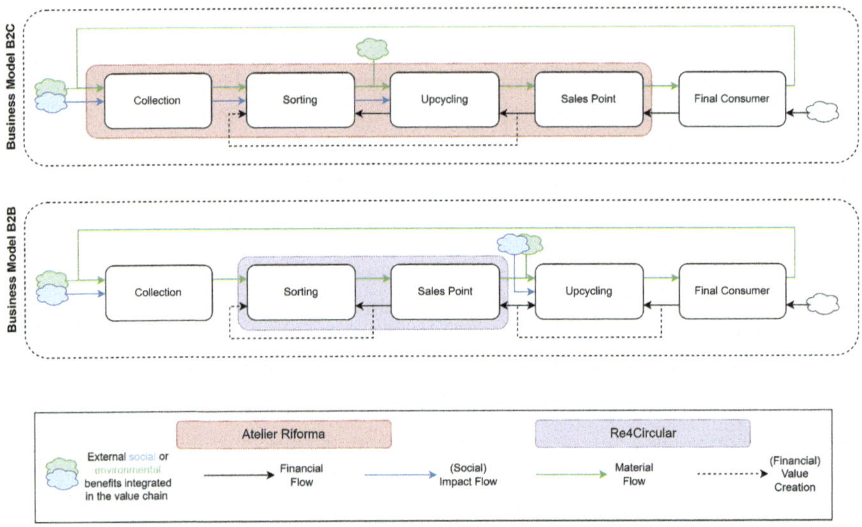

Fig. 3 Financial material impact flow analysis. (Source: Authors' elaboration)

start-up generates its revenues by taking a margin on the economic benefits it produces for sellers and buyers through its technology and online marketplace.

By examining these flows, we can see that Atelier Riforma's business models both contribute to reducing environmental impact and promoting social inclusion, though in different ways. The B2C model involves more direct engagement with consumers and internal generation of social impact, while the B2B model leverages technology to scale these impacts through partnerships. The financial sustainability of the start-up is maintained in both models by capturing economic value either directly from consumers or indirectly through a margin on transactions facilitated by the platform.

5 Discussion

5.1 Atelier Riforma's Social and Environmental Impacts

Regarding environmental impacts, the main contribution of Atelier Riforma in reducing environmental impacts in the fashion and textile sector is waste avoidance and potentially reduced exploitation of natural resources. As the circular textile business expands, less textile waste is generated, and fewer resources are depleted for new textile production, assuming that average consumption levels do not increase in the transition from linear to circular goods, as assumed by studies such as Saha et al. (2021). This reduction in environmental impact is reflected in the indicators selected to monitor the used clothes and deadstock fabrics cataloged and uploaded on the platform, which measure waste avoidance (Shirvanimoghaddam et al., 2020), and by measuring the CO_2 emissions avoided by promoting short supply chains within the local circular market, as opposed to traditional linear garments transported across global value chains (Chen et al., 2021). The financial material impact flow analysis captures how environmental impacts shift with the modification of the business model. Specifically, the Re4Circular technology contributes to upcycling and reuse, even without physically managing the material, thereby externalizing (but promoting) a portion of the positive environmental impacts of the business. Given the embeddedness of the start-up's environmental mission, Atelier Riforma qualifies as a mission-centric enterprise according to Alter's (2007) social enterprise classification. This means that the enterprise is central to the organization's environmental mission.

The social dimension of Atelier Riforma's impact is substantial and multifaceted. The company's efforts to promote good working conditions for disadvantaged groups directly address significant social externalities in the traditional linear model of the textile industry, such as insufficient wages, health and safety issues, child labor, and forced labor (Desore & Narula, 2018). This aligns with broader literature emphasizing the need for ethical labor practices in global supply chains. Additionally, Atelier Riforma's public information and dissemination activities on responsible consumption resonate with the goals of sustainable development (Geiger et al., 2018) and the United Nations Sustainable Development Goal 12 (SDG 12), which advocates for responsible consumption and production. Moreover, promoting responsible consumption through continuous public engagement empowers consumers, transforming them from mere buyers to active citizens who make informed decisions about their purchases and their environmental impact (Fischer et al., 2021; Hertwig & Grüne-Yanoff, 2017). Considering that Atelier Riforma also provides its technology free of charge to nonprofit organizations that collect used clothes for solidarity purposes, its social impact broadens to encompass a wider empowerment of the local community. This social value extends beyond providing increased new employment opportunities for disadvantaged people to encompass political, economic, social, emotional, creative, and self-development domains (Tanekenov et al., 2018), promoting a socially responsible circular community.

We argue that these aspects qualify the social mission of Atelier Riforma as an important part of its business, and by adopting Alter's (2007) social enterprise classification, it can be considered at least mission-related. A final consideration is that even though the start-up has automatic social value spillovers occurring alongside its key operations, it does not deliver its social value directly through its products to its clients. According to Santos et al. (2015), this may result in an intermediate risk of mission drift and a moderately difficult path to financial sustainability. However, this is only valid for the social dimension of the value spillovers, as the environmental value is fully embedded in its business operation. This comprehensive approach to social impact underscores the company's role in fostering a socially inclusive and environmentally sustainable fashion industry.

5.2 *Toward a Social Circular Business Model and Its Assessment*

From the results and discussion of Atelier Riforma's social and environmental impacts, it is clear that these dimensions are fully present and coexist in the business model. This was highlighted during the interview with the CEO, who stated:

> *The environmental and social dimensions are closely integrated. At times, the emphasis may shift more towards one aspect or the other depending on the circumstances, but generally, Atelier Riforma's social and environmental initiatives support and reinforce each other. This intersection is fundamental for the sustainability of our circularity-based business model.*

The integration of these aspects is also evident in the key activities mapped in the social circular business model canvas, which show that many business initiatives, such as the promotion of upcycling through socially responsible organizations in the textile sector, and the communication, sensitization, and activist roles to empower the community through promoting circularity practices, are intertwined. This integration also emerges in the financial material impact flow analysis, which maps the social and environmental values through the key steps of Atelier Riforma's value chain.

Based on these premises, we claim that Atelier Riforma constitutes a case of an integrated social circular business model in the textile sector. Beyond its key circular operation of recovering garments that would otherwise become waste, the start-up responds to the final consumer's needs for transparency and valorization of virtuous organizations in this sector. This activity has the potential to significantly influence the development of apparel firms' business models, pushing companies toward a circular model (Abbate et al., 2023). However, it should be noted that despite multiple studies addressing CBMs in the textile, apparel, and fashion sector (Salmi & Kaipia, 2022), the conceptualization and potential contribution to sustainability of social circular business models in this industry have not yet been investigated in the literature. Although studies like those by Hultberg and Pal (2021), who have examined strategies for scaling CBMs in this sector, and Oliveira Neto et al. (2024),who have identified circular practices for executing new projects that meet customer needs, the potential implications of social circular business models have not yet been directly tackled.

The relevance and integration of social and circular aspects are evident from our results, but assessing, measuring, and accounting for the added value derived from this integration are challenging. The business model canvas shows key activities and maps specific operations where social and environmental spillover values emerge, but it is not a useful tool for impact assessment (Daou et al., 2020). Through their theory of change analysis, Atelier Riforma has identified social and environmental indicators to assess their impact. However, as the CEO stated, assessing and measuring the value derived from the integration and interrelation of the social and environmental aspects remain very challenging, even with the theory of change approach. This difficulty is reflected in the lack of research on social circular business models and the absence of methods and tools to guide this type of assessment (Millar et al., 2019).

5.3 Limitations and Future Research

A first limitation of our study is that the adoption of a single-case study methodology limits the contribution of this study to the specific context examined (Vaivio, 2008). The insights derived from Atelier Riforma's experience are confined to the textile sector, even though social circular business models may exist in other sectors. Additionally, our analysis has not been longitudinal; tracking the evolution of the business over time could provide new insights into the development of a social circular business model in the textile sector. Furthermore, a suggestion for future studies is to incorporate a comparative analysis of multiple cases to enhance the generalizability of the findings.

Future research should focus on further conceptualizing social circular business models, particularly to understand their presence and main differences across various sectors. It is crucial to explore integrated impact assessment, financial sustainability, scalability, and contributions to sustainable development within this framework. Our study focuses on the microlevel; thus, future studies could expand the scope of analysis to the meso-level to fully capture the empirical emergence of integrated social and environmental circularity models in the expanding circular textile sector (Pitkänen et al., 2023). Finally, despite the identification of social impacts related to community development and empowerment that emerged from our data, we were unable to fully capture and assess these impacts in our financial material impact flow analysis, as it is limited to the main steps of Atelier Riforma's value chain. Future studies should develop methods to comprehensively assess and account for all the integrated and intertwined social and environmental implications derived from these kinds of social circular business activities and the beneficial effects of integrating these two aspects.

6 Conclusion

This study illuminates the integration of social and environmental dimensions within circular business models through the detailed case analysis of Atelier Riforma. By examining this innovative start-up, we have demonstrated how such models can simultaneously achieve environmental sustainability and social equity, addressing significant challenges in the textile and fashion industry. The environmental benefits are evident through waste reduction, resource efficiency, and increased transparency in the textile and fashion circular sector promoted by the Re4Circular technology, while the social impacts include improved labor conditions and community empowerment.

Atelier Riforma's business model showcases the potential of integrating CE principles with a strong social mission. The case study reveals that the start-up's activities, from promoting upcycling to supporting disadvantaged groups, create a synergistic effect that enhances both social and environmental outcomes. This dual impact is essential for fostering a sustainable and inclusive fashion industry. However, the challenge remains in accurately assessing and measuring these intertwined impacts. Traditional tools like the business model canvas and theory of change provide frameworks, but there is a need for more sophisticated methods to capture the full spectrum of benefits derived from the interaction of social and environmental value spillovers and their interrelation in circular business models.

Our findings suggest that Atelier Riforma serves as a valuable model for other sectors looking to adopt similar integrated approaches. The start-up's evolution from a labor-intensive B2C model to a scalable B2B platform powered by AI highlights the importance of innovation and adaptability in achieving sustainability goals. The use of technology not only improves efficiency but also expands the potential for broader impact by connecting various stakeholders in the circular fashion value chain.

Despite the promising results, this study acknowledges several limitations. The single-case study approach confines the findings to the specific context of the textile sector, and the lack of longitudinal analysis limits our understanding of the long-term impacts. Future research should explore multiple cases across different sectors to generalize the findings and provide a comprehensive understanding of social circular business models. Additionally, developing robust methods for impact assessment remains a crucial area for further investigation.

The implications of this study are significant for researchers, practitioners, and policymakers. It underscores the need for continued development of SCBMs and their integration into various industries. Policymakers, in particular, can play a pivotal role by creating supportive regulations and incentives that promote the adoption and scalability of these models. For practitioners, the insights from Atelier Riforma offer practical guidance on implementing and scaling social circular initiatives.

This study contributes to the growing body of literature on social circular business models by providing a detailed case analysis of Atelier Riforma. It highlights the potential of these models to address dual objectives of environmental sustainability and social equity through circularity principles, offering valuable insights for fostering sustainable development through innovative business practices. The findings underscore the need for continued research and development of methodologies to fully assess and harness the benefits of social circular business models in various industries. The journey of Atelier Riforma exemplifies the transformative potential of integrating circular economy principles with social missions, paving the way for a more sustainable and equitable future in the fashion industry and beyond.

References

Abbate, S., Centobelli, P., & Cerchione, R. (2023). From fast to slow: An exploratory analysis of circular business models in the Italian apparel industry. *International Journal of Production Economics, 260*, 108824. https://doi.org/10.1016/j.ijpe.2023.108824

Abdelmeguid, A., Afy-Shararah, M., & Salonitis, K. (2024). Towards circular fashion: Management strategies promoting circular behaviour along the value chain. *Sustainable Production and Consumption, 48*, 143–156. https://doi.org/10.1016/j.spc.2024.05.010

Abela, M. (2022). A new direction? The 'mainstreaming' of sustainability reporting. *Sustainability Accounting, Management and Policy Journal, 13*, 1261–1283. https://doi.org/10.1108/SAMPJ-06-2021-0201

Alter, K. (2007). Social enterprise typology. *Virtue Ventures LLC, 12*, 1–124.

Anderson, M. (2018). Fair trade and consumer social responsibility. *Management Decision, 56*, 634–651. https://doi.org/10.1108/MD-01-2017-0013

Arribas-Ibar, M., Nylund, P. A., & Brem, A. (2022). Circular business models in the luxury fashion industry: Toward an ecosystemic dominant design? *Current Opinion in Green and Sustainable Chemistry, 37*, 100673. https://doi.org/10.1016/j.cogsc.2022.100673

Assmann, I. R., Rosati, F., & Morioka, S. N. (2023). Determinants of circular business model adoption – A systematic literature review. *Business Strategy and the Environment, 32*, 6008–6028. https://doi.org/10.1002/bse.3470

Aureli, S., Foschi, E., & Paletta, A. (2023). Management accounting for a circular economy: Current limits and avenue for a dialogic approach. *Accounting, Auditing & Accountability Journal*. https://doi.org/10.1108/AAAJ-04-2022-5766

Barletta, M., D'Adamo, I., Garza-Reyes, J. A., & Gastaldi, M. (2024). Business strategy and innovative models in the fashion industry: Clothing leasing as a driver of sustainability. *Business Strategy and the Environment*. https://doi.org/10.1002/bse.3723

Beyer, K., & Arnold, M. G. (2022). Examining the social side of sustainability in the debate on business model innovations in the textile, clothing and fashion industry: A typology based on the value chain perspective. *International Journal of Innovation and Sustainable Development, 16*, 322–371. https://doi.org/10.1504/IJISD.2022.123908

Bowen, G. A. (2009). Document analysis as a qualitative research method. *Qualitative Research Journal, 9*, 27–40. https://doi.org/10.3316/QRJ0902027

Braungart, M., McDonough, W., & Bollinger, A. (2007). Cradle-to-cradle design: Creating healthy emissions – A strategy for eco-effective product and system design. *Journal of Cleaner Production, 15*, 1337–1348. https://doi.org/10.1016/j.jclepro.2006.08.003

Brundtland, G. H., & Khalid, M. (1987). *Our common future*. Oxford University Press.

Bubicz, M. E., Dias Barbosa-Póvoa, A. P. F., & Carvalho, A. (2021). Social sustainability management in the apparel supply chains. *Journal of Cleaner Production, 280*, 124214. https://doi.org/10.1016/j.jclepro.2020.124214

C2C Institute. (2021). Cradle to Cradle Certified Version 4.0 Product Standard.

Calisto Friant, M., Vermeulen, W. J. V., & Salomone, R. (2020). A typology of circular economy discourses: Navigating the diverse visions of a contested paradigm. *Resources, Conservation and Recycling, 161*, 104917. https://doi.org/10.1016/j.resconrec.2020.104917

Chen, X., Memon, H. A., Wang, Y., Marriam, I., & Tebyetekerwa, M. (2021). Circular economy and sustainability of the clothing and textile industry. *Materials Circular Economy, 3*, 12. https://doi.org/10.1007/s42824-021-00026-2

Clark, W. H., Jr., & Babson, E. K. (2012). How benefit corporations are redefining the purpose of business corporations. *William Mitchell Law Rev, 38*, 817–850.

Clube, R. K. M., & Tennant, M. (2020). The circular economy and human needs satisfaction: Promising the radical, delivering the familiar. *Ecological Economics, 177*, 106772. https://doi.org/10.1016/j.ecolecon.2020.106772

Connell, J. P., & Kubisch, A. C. (1998). Applying a theory of change approach to the evaluation of comprehensive community initiatives: Progress, prospects, and problems. In

K. Fulbright-Anderson, A. C. Kubisch, & J. P. Connell (Eds.), *New approaches to evaluating community initiatives* (pp. 1–16). The Aspen Institute.
Cooper, D. J., & Morgan, W. (2008). Case study research in accounting. *Accounting Horizons, 22*, 159–178. https://doi.org/10.2308/acch.2008.22.2.159
D'Adamo, I., Lupi, G., Morone, P., & Settembre-Blundo, D. (2022). Towards the circular economy in the fashion industry: The second-hand market as a best practice of sustainable responsibility for businesses and consumers. *Environmental Science and Pollution Research, 29*, 46620–46633. https://doi.org/10.1007/s11356-022-19255-2
Dainelli, F., Daddi, T., & Marrucci, L. (2024). Financial sustainability of circular innovations in SMEs. A case study from the fashion industry in Italy. *Journal of Cleaner Production, 451*, 142042. https://doi.org/10.1016/j.jclepro.2024.142042
Dale, S., Frost, M., & Ison, S. (2023). The theory of change and realistic evaluation applied to the evaluation of a transport intervention: The case of the Nottingham workplace parking levy. *Evaluation and Program Planning, 98*, 102282. https://doi.org/10.1016/j.evalprogplan.2023.102282
Daou, A., Mallat, C., Chammas, G., Cerantola, N., Kayed, S., & Saliba, N. A. (2020). The Ecocanvas as a business model canvas for a circular economy. *Journal of Cleaner Production, 258*, 120938. https://doi.org/10.1016/j.jclepro.2020.120938
de Oliveira Neto, G. C., Santos, R. A. R., Pinto, L. F. R., Flausino, F. R., de Oliveira, D. E. P., & Seri, M. N. (2024). Innovative circular practices integrating business model for textile industry. *Journal of Engineered Fibers and Fabrics, 19*. https://doi.org/10.1177/15589250241226481
Deckers, J., Duhoux, T., Due, S., 2024. Textile waste management in Europe's circular economy.
Desore, A., & Narula, S. A. (2018). An overview on corporate response towards sustainability issues in textile industry. *Environment, Development and Sustainability, 20*, 1439–1459. https://doi.org/10.1007/s10668-017-9949-1
Dey, P. K., Malesios, C., De, D., Budhwar, P., Chowdhury, S., & Cheffi, W. (2020). Circular economy to enhance sustainability of small and medium-sized enterprises. *Business Strategy and the Environment, 29*, 2145–2169. https://doi.org/10.1002/bse.2492
Dyer, W. G., & Wilkins, A. L. (1991). Better stories, not better constructs, to generate better theory: A rejoinder to Eisenhardt. *Academy of Management Review, 16*, 613–619. https://doi.org/10.5465/amr.1991.4279492
Eisenhardt, K. M., & Graebner, M. E. (2007). Theory building from cases: Opportunities and challenges. *The Academy of Management Journal, 50*, 25–32.
Elkington, J. (1998). Partnerships from cannibals with forks: The triple bottom line of 21st-century business. *Environmental Quality Management, 8*, 37–51. https://doi.org/10.1002/tqem.3310080106
Fehrer, J. A., & Wieland, H. (2021). A systemic logic for circular business models. *Journal of Business Research, 125*, 609–620. https://doi.org/10.1016/j.jbusres.2020.02.010
Fischer, D., Reinermann, J.-L., Guillen Mandujano, G., DesRoches, C. T., Diddi, S., & Vergragt, P. J. (2021). Sustainable consumption communication: A review of an emerging field of research. *Journal of Cleaner Production, 300*, 126880. https://doi.org/10.1016/j.jclepro.2021.126880
Fletcher, K. (2010). Slow fashion: An invitation for systems change. *Fashion Practice, 2*, 259–265. https://doi.org/10.2752/175693810X12774625387594
García-Muiña, F., Medina-Salgado, M. S., González-Sánchez, R., Huertas-Valdivia, I., Ferrari, A. M., & Settembre-Blundo, D. (2021). Industry 4.0-based dynamic social organizational life cycle assessment to target the social circular economy in manufacturing. *Journal of Cleaner Production, 327*, 129439. https://doi.org/10.1016/j.jclepro.2021.129439
Geiger, S. M., Fischer, D., & Schrader, U. (2018). Measuring what matters in sustainable consumption: An integrative framework for the selection of relevant behaviors. *Sustainable Development, 26*, 18–33. https://doi.org/10.1002/sd.1688
Geissdoerfer, M., Savaget, P., Bocken, N. M. P., & Hultink, E. J. (2017). The circular economy – A new sustainability paradigm? *Journal of Cleaner Production, 143*, 757–768. https://doi.org/10.1016/j.jclepro.2016.12.048

Geissdoerfer, M., Pieroni, M. P. P., Pigosso, D. C. A., & Soufani, K. (2020). Circular business models: A review. *Journal of Cleaner Production, 277*, 123741. https://doi.org/10.1016/j.jclepro.2020.123741

Ghisellini, P., Cialani, C., & Ulgiati, S. (2016). A review on circular economy: The expected transition to a balanced interplay of environmental and economic systems. *Journal of Cleaner Production, 114*, 11–32. https://doi.org/10.1016/j.jclepro.2015.09.007

Guibrunet, L., Sanzana Calvet, M., & Castán Broto, V. (2017). Flows, system boundaries and the politics of urban metabolism: Waste management in Mexico City and Santiago de Chile. *Geoforum, 85*, 353–367. https://doi.org/10.1016/j.geoforum.2016.10.011

Halkos, G., & Nomikos, S. (2021). Corporate social responsibility: Trends in global reporting initiative standards. *Economic Analysis and Policy, 69*, 106–117. https://doi.org/10.1016/j.eap.2020.11.008

Hertwig, R., & Grüne-Yanoff, T. (2017). Nudging and boosting: Steering or empowering good decisions. *Perspectives on Psychological Science, 12*, 973–986. https://doi.org/10.1177/1745691617702496

Hidalgo-Carvajal, D., Carrasco-Gallego, R., & Morales-Alonso, G. (2021). From goods to services and from linear to circular: The role of servitization's challenges and drivers in the shifting process. *Sustainability, 13*, 4539. https://doi.org/10.3390/su13084539

Huang, C.-L., Vause, J., Ma, H.-W., & Yu, C.-P. (2012). Using material/substance flow analysis to support sustainable development assessment: A literature review and outlook. *Resources, Conservation and Recycling, 68*, 104–116. https://doi.org/10.1016/j.resconrec.2012.08.012

Hultberg, E., & Pal, R. (2021). Lessons on business model scalability for circular economy in the fashion retail value chain: Towards a conceptual model. *Sustainable Production and Consumption, 28*, 686–698. https://doi.org/10.1016/j.spc.2021.06.033

Impact Institute. (2019). *The true price of jeans.*

Kanzari, A., Rasmussen, J., Nehler, H., & Ingelsson, F. (2022). How financial performance is addressed in light of the transition to circular business models – A systematic literature review. *Journal of Cleaner Production, 376*, 134134. https://doi.org/10.1016/j.jclepro.2022.134134

Kevin van Langen, S., Vassillo, C., Ghisellini, P., Restaino, D., Passaro, R., & Ulgiati, S. (2021). Promoting circular economy transition: A study about perceptions and awareness by different stakeholders groups. *Journal of Cleaner Production, 316*, 128166. https://doi.org/10.1016/j.jclepro.2021.128166

Kirchherr, J. (2021). Towards circular justice: A proposition. *Resources, Conservation and Recycling, 173*, 105712. https://doi.org/10.1016/j.resconrec.2021.105712

Kirchherr, J., Reike, D., & Hekkert, M. (2017). Conceptualizing the circular economy: An analysis of 114 definitions. *Resources, Conservation and Recycling, 127*, 221–232. https://doi.org/10.1016/j.resconrec.2017.09.005

Köksal, D., Strähle, J., Müller, M., & Freise, M. (2017). Social sustainable supply chain management in the textile and apparel industry – A literature review. *Sustainability, 9*, 100. https://doi.org/10.3390/su9010100

Korhonen, J., Honkasalo, A., & Seppälä, J. (2018). Circular economy: The concept and its limitations. *Ecological Economics, 143*, 37–46. https://doi.org/10.1016/j.ecolecon.2017.06.041

Lewandowski, M. (2016). Designing the business models for circular economy – Towards the conceptual framework. *Sustainability, 8*, 43. https://doi.org/10.3390/su8010043

Llorach-Massana, P., Farreny, R., & Oliver-Solà, J. (2015). Are cradle to cradle certified products environmentally preferable? Analysis from an LCA approach. *Journal of Cleaner Production, 93*, 243–250. https://doi.org/10.1016/j.jclepro.2015.01.032

Lüdeke-Freund, F., Froese, T., Dembek, K., Rosati, F., & Massa, L. (2024). What makes a business model sustainable? Activities, design themes, and value functions. *Organization and Environment*. https://doi.org/10.1177/10860266241235212

Margoluis, R., Stem, C., Salafsky, N., & Brown, M. (2009). Using conceptual models as a planning and evaluation tool in conservation. *Evaluation and Program Planning, 32*, 138–147. https://doi.org/10.1016/j.evalprogplan.2008.09.007

Masi, D., Day, S., & Godsell, J. (2017). Supply chain configurations in the circular economy: A systematic literature review. *Sustainability, 9*, 1602. https://doi.org/10.3390/su9091602

Mattos, C. A., Scur, G., & Albuquerque, T. L. M. (2022). Evaluation of circular business model: Theory of change approach. *Evaluation and Program Planning, 92*, 102069. https://doi.org/10.1016/j.evalprogplan.2022.102069

Mayne, J. (2015). Useful theory of change models. *Canadian Journal of Program Evaluation, 30*, 119–142. https://doi.org/10.3138/cjpe.230

Mayne, J. (2017). Theory of change analysis: Building robust theories of change. *Canadian Journal of Program Evaluation, 32*, 155–173. https://doi.org/10.3138/cjpe.31122

Merli, R., Preziosi, M., & Acampora, A. (2018). How do scholars approach the circular economy? A systematic literature review. *Journal of Cleaner Production, 178*, 703–722. https://doi.org/10.1016/j.jclepro.2017.12.112

Michelini, L., Grieco, C., Ciulli, F., & Di Leo, A. (2020). Uncovering the impact of food sharing platform business models: A theory of change approach. *British Food Journal, 122*, 1437–1462. https://doi.org/10.1108/BFJ-06-2019-0422

Mies, A., & Gold, S. (2021). Mapping the social dimension of the circular economy. *Journal of Cleaner Production, 321*, 128960. https://doi.org/10.1016/j.jclepro.2021.128960

Millar, N., McLaughlin, E., & Börger, T. (2019). The circular economy: Swings and roundabouts? *Ecological Economics, 158*, 11–19. https://doi.org/10.1016/j.ecolecon.2018.12.012

Moreno, M., De los Rios, C., Rowe, Z., & Charnley, F. (2016). A conceptual framework for circular design. *Sustainability, 8*, 937. https://doi.org/10.3390/su8090937

Muñoz, P., Cacciotti, G., & Cohen, B. (2018). The double-edged sword of purpose-driven behavior in sustainable venturing. *Journal of Business Venturing, 33*, 149–178. https://doi.org/10.1016/j.jbusvent.2017.12.005

Naudé, M. (2011). Sustainable development in companies: Theoretical dream or implementable reality? *Corporate Ownership and Control, 8*, 352–364. https://doi.org/10.22495/cocv8i4c3art4

Nayak, R., Van Thang, L. N., Nguyen, T., Gaimster, J., Morris, R., & George, M. (2022). Sustainable developments and corporate social responsibility in Vietnamese fashion enterprises. *Journal of Fashion Marketing and Management: An International Journal, 26*, 307–327. https://doi.org/10.1108/JFMM-07-2020-0148

Niero, M., Olsen, S. I., & Laurent, A. (2018). Renewable energy and carbon management in the Cradle-to-Cradle Certification: Limitations and opportunities. *Journal of Industrial Ecology, 22*, 760–772. https://doi.org/10.1111/jiec.12594

Nigri, G., & Del Baldo, M. (2018). Sustainability reporting and performance measurement systems: How do small- and medium-sized benefit corporations manage integration? *Sustainability, 10*, 4499. https://doi.org/10.3390/su10124499

Opferkuch, K., Caeiro, S., Salomone, R., & Ramos, T. B. (2021). Circular economy in corporate sustainability reporting: A review of organisational approaches. *Business Strategy and the Environment, 30*, 4015–4036. https://doi.org/10.1002/bse.2854

Padilla-Rivera, A., do Carmo, B. B. T., Arcese, G., & Merveille, N. (2021). Social circular economy indicators: Selection through fuzzy delphi method. *Sustainable Production and Consumption, 26*, 101–110. https://doi.org/10.1016/j.spc.2020.09.015

Peters, G., Li, M., & Lenzen, M. (2021). The need to decelerate fast fashion in a hot climate – A global sustainability perspective on the garment industry. *Journal of Cleaner Production, 295*, 126390. https://doi.org/10.1016/j.jclepro.2021.126390

Pitkänen, K., Karppinen, T. K. M., Kautto, P., Pirtonen, H., Salmenperä, H., Savolahti, H., Schubin, E., & Myllymaa, T. (2023). How to measure the social sustainability of the circular economy? Developing and piloting social circular economy indicators in Finland. *Journal of Cleaner Production, 392*, 136238. https://doi.org/10.1016/J.JCLEPRO.2023.136238

Puntillo, P., Gulluscio, C., Huisingh, D., & Veltri, S. (2021). Reevaluating waste as a resource under a circular economy approach from a system perspective: Findings from a case study. *Business Strategy and the Environment, 30*, 968–984. https://doi.org/10.1002/bse.2664

Qu, S. Q., & Dumay, J. (2011). The qualitative research interview. *Qualitative Research in Accounting & Management, 8*, 238–264. https://doi.org/10.1108/11766091111162070

Robinson, S. (2017). *Social circular economy: Opportunities for people planet and profit*. Social Circular Economy.

Rosa, P., Sassanelli, C., & Terzi, S. (2019). Towards circular business models: A systematic literature review on classification frameworks and archetypes. *Journal of Cleaner Production, 236*, 117696. https://doi.org/10.1016/j.jclepro.2019.117696

Rossi, A., & Luque-Vílchez, M. (2021). The implementation of sustainability reporting in a small and medium enterprise and the emergence of integrated thinking. *Meditari Accountancy Research, 29*, 966–984. https://doi.org/10.1108/MEDAR-02-2020-0706

Saha, K., Dey, P. K., & Papagiannaki, E. (2021). Implementing circular economy in the textile and clothing industry. *Business Strategy and the Environment, 30*, 1497–1530. https://doi.org/10.1002/bse.2670

Saidani, M., Shevchenko, T., Esfandabadi, Z. S., Ranjbari, M., Mesa, J. A., Yannou, B., & Cluzel, F. (2024). The future of circular economy metrics: Expert visions. *Resources, Conservation and Recycling, 205*, 107565. https://doi.org/10.1016/j.resconrec.2024.107565

Salmi, A., & Kaipia, R. (2022). Implementing circular business models in the textile and clothing industry. *Journal of Cleaner Production, 378*, 134492. https://doi.org/10.1016/j.jclepro.2022.134492

Salvador, R., Barros, M. V., da Luz, L. M., Piekarski, C. M., & de Francisco, A. C. (2020). Circular business models: Current aspects that influence implementation and unaddressed subjects. *Journal of Cleaner Production, 250*, 119555. https://doi.org/10.1016/j.jclepro.2019.119555

Santos, F., Pache, A.-C., & Birkholz, C. (2015). Making hybrids work: Aligning business models and organizational design for social enterprises. *California Management Review, 57*, 36–58. https://doi.org/10.1525/cmr.2015.57.3.36

Scarpellini, S. (2022). Social impacts of a circular business model: An approach from a sustainability accounting and reporting perspective. *Corporate Social Responsibility and Environmental Management, 29*, 646–656. https://doi.org/10.1002/csr.2226

Scarpellini, S., Valero-Gil, J., Moneva, J. M., & Andreaus, M. (2020). Environmental management capabilities for a "circular eco-innovation.". *Business Strategy and the Environment, 29*, 1850–1864. https://doi.org/10.1002/bse.2472

Schroeder, P., Anggraeni, K., & Weber, U. (2019). The relevance of circular economy practices to the sustainable development goals. *Journal of Industrial Ecology, 23*, 77–95. https://doi.org/10.1111/jiec.12732

Schwandt, T. (2015). *Evaluation foundations revisited: Cultivating a life of the mind for practice*. Stanford University Press.

Selvefors, A., Renström, S., Whalen, K. A., Fallahi, S., Leivas, M., Nordenö, H., & Fransson, A. (2024). User-centered circular value propositions – Approaches in practice and research. *Resources, Conservation and Recycling, 207*, 107628. https://doi.org/10.1016/j.resconrec.2024.107628

Shields, J. F., & Shelleman, J. M. (2017). A method to launch sustainability reporting in SMEs: The B Corp impact assessment framework. *Journal of Strategic Innovation and Sustainability, 12*. https://doi.org/10.33423/jsis.v12i2.798

Shirvanimoghaddam, K., Motamed, B., Ramakrishna, S., & Naebe, M. (2020). Death by waste: Fashion and textile circular economy case. *Science of the Total Environment, 718*, 137317. https://doi.org/10.1016/j.scitotenv.2020.137317

Silva, M. E., Ashby, A., & Nascimento, C. M. (2023). Social sustainability in micro and small enterprise supply chains: Empirical insights from the clothing industry. *Latin American Business Review, 24*, 1–30. https://doi.org/10.1080/10978526.2021.1992597

Solis, M., Huygens, D., Tonini, D., & Fruergaard Astrup, T. (2024). Management of textile waste in Europe: An environmental and a socio-economic assessment of current and future scenarios. *Resources, Conservation and Recycling, 207*, 107693. https://doi.org/10.1016/j.resconrec.2024.107693

Souza Piao, R., de Vincenzi, T. B., da Silva, A. L. F., de Oliveira, M. C. C., Vazquez-Brust, D., & Monteiro Carvalho, M. (2023). How is the circular economy embracing social inclusion? *Journal of Cleaner Production, 411*, 137340. https://doi.org/10.1016/J.JCLEPRO.2023.137340

Suárez-Eiroa, B., Fernández, E., Méndez-Martínez, G., & Soto-Oñate, D. (2019). Operational principles of circular economy for sustainable development: Linking theory and practice. *Journal of Cleaner Production, 214*, 952–961. https://doi.org/10.1016/j.jclepro.2018.12.271

Susur, E., & Engwall, M. (2023). A transitions framework for circular business models. *Journal of Industrial Ecology, 27*, 19–32. https://doi.org/10.1111/jiec.13363

Tanekenov, A., Fitzpatrick, S., & Johnsen, S. (2018). Empowerment, capabilities and homelessness: The limitations of employment-focused social enterprises in addressing complex needs. *Housing, Theory and Society, 35*, 137–155. https://doi.org/10.1080/14036096.2017.1290676

Thorisdottir, T. S., & Johannsdottir, L. (2020). Corporate social responsibility influencing sustainability within the fashion industry. A systematic review. *Sustainability, 12*, 9167. https://doi.org/10.3390/su12219167

Todeschini, B. V., Cortimiglia, M. N., Callegaro-de-Menezes, D., & Ghezzi, A. (2017). Innovative and sustainable business models in the fashion industry: Entrepreneurial drivers, opportunities, and challenges. *Business Horizons, 60*, 759–770. https://doi.org/10.1016/j.bushor.2017.07.003

Ünal, E., Urbinati, A., & Chiaroni, D. (2019). Managerial practices for designing circular economy business models. *Journal of Manufacturing Technology Management, 30*, 561–589. https://doi.org/10.1108/JMTM-02-2018-0061

Vahidzadeh, R., Bertanza, G., Sbaffoni, S., & Vaccari, M. (2021). Regional industrial symbiosis: A review based on social network analysis. *Journal of Cleaner Production, 280*, 124054. https://doi.org/10.1016/j.jclepro.2020.124054

Vaivio, J. (2008). Qualitative management accounting research: Rationale, pitfalls and potential. *Qualitative Research in Accounting & Management, 5*, 64–86. https://doi.org/10.1108/11766090810856787

Vanhuyse, F., Fejzić, E., Ddiba, D., & Henrysson, M. (2021). The lack of social impact considerations in transitioning towards urban circular economies: A scoping review. *Sustainable Cities and Society, 75*, 103394. https://doi.org/10.1016/j.scs.2021.103394

Voss, C., Tsikriktsis, N., & Frohlich, M. (2002). Case research in operations management. *International Journal of Operations & Production Management, 22*, 195–219. https://doi.org/10.1108/01443570210414329

Walker, A. M., Opferkuch, K., Roos Lindgreen, E., Simboli, A., Vermeulen, W. J. V., & Raggi, A. (2021). Assessing the social sustainability of circular economy practices: Industry perspectives from Italy and The Netherlands. *Sustainable Production and Consumption, 27*, 831–844. https://doi.org/10.1016/j.spc.2021.01.030

Warde, A. (2015). The sociology of consumption: Its recent development. *Annual Review of Sociology, 41*, 117–134. https://doi.org/10.1146/annurev-soc-071913-043208

Yin, R. K. (2018). *Case study research and applications*. Sage.

Rise of Circular Fashion for Sustainable Apparel Retailing

R. Surjit, R. Akshatha, R. Akshya, and C. Harani

1 Introduction

Circular fashion is a concept within the broader circular economy framework. It focuses on reducing waste and pollution by maintaining the products in some form always avoiding its recycling so that waste is minimised and the pollution burden is reduced. It aims to shift the traditional linear fashion model where products are made, used, and discarded to a more sustainable model which leads to extending the product life cycle by reusing them. Several academic papers and reports have explored the various business models that can support this transition. The growth of circular fashion can rise steeply by increasing customer demand for sustainability, technological advancements, and economic benefits. There are also subscription services, reselling, and renting models reshaping the industry by offering sustainable alternatives to ownership, though they face challenges in quality control and consumer fit. Consumer behaviour increasingly drives the shift towards sustainable fashion, emphasising the need for transparency and engagement. Addressing challenges like greenwashing and affordability, and leveraging opportunities for increased consumer power and supportive policies, is crucial for advancing a circular fashion ecosystem. Overall, circular fashion represents a significant step towards achieving sustainability goals in the industry. The shift from linear to circular fashion focused on extending product life and reducing waste, demanding strong government action and global collaboration. Policies like Extended Producer Responsibility (EPR), green procurement, and eco-labelling address market failures

R. Surjit (✉) · R. Akshatha · R. Akshya · C. Harani
Department of Fashion Technology, PSG College of Technology,
Coimbatore, Tamil Nadu, India
e-mail: rst.fashion@psgtech.ac.in; 22h202@psgtech.ac.in; 22h203@psgtech.ac.in; 22h207@psgtech.ac.in

S. S. Muthu (ed.), *Circular Business Models in the Apparel Industry*,
Environmental Footprints and Eco-design of Products and Processes,
https://doi.org/10.1007/978-3-031-80193-8_6

and encourage circular practices. Global efforts, such as those by the UN Alliance for Sustainable Fashion and the Ellen MacArthur Foundation, foster sustainability through international cooperation. Circular fashion economically benefits businesses through cost savings, new revenue streams, and enhanced brand image, while consumers gain access to affordable and sustainable fashion. Technology, including blockchain and digital tools, supports this transition by improving efficiency and engagement. Collaboration among governments, businesses, and consumers, supported by technological advances, is essential for a sustainable fashion industry. Supply chain management and transparency are essential for circular fashion in apparel retail, involving life cycle oversight, sustainable practices, and social responsibility. Transparency builds trust and improves efficiency but faces challenges like complexity and cost. Technological advancements, including blockchain, Internet of Things (IoT), and artificial intelligence (AI), enhance supply chain visibility but encounter integration and cybersecurity issues. Barriers to circular fashion include consumer behaviour and regulatory gaps. Future success relies on technology investment, collaboration, and scaling recycling infrastructure to promote sustainability in the fashion industry.

2 Circular Business Models

The various models that are used in 'circular fashion' are discussed below:

- Closed-Loop Recycling and Upcycling

  Closed-loop recycling is a model where waste materials from one process are collected, reprocessed, and used to create new products, ideally with minimal loss in quality. Upcycling, on the other hand, involves creating new products of higher quality or value from waste materials. These models help reduce resource extraction and waste generation.

  According to Bocken et al. (2016), closed-loop recycling can create new revenue streams from waste materials, while upcycling adds value to products that would otherwise be discarded. Similarly, Kirchherr et al. (2017) discuss how these models align with the principles of a circular economy by reducing environmental impacts and fostering sustainable consumption.

- Products-as-a-Service (PaaS)

  The products-as-a-service model is a very effective way of utilising the products repeatedly without throwing the product as waste as the material finds usage every time somewhere thereby avoiding landfills. This is a similar model of SaaS and PaaS used in the software industry, which is used here, where the products are given for rentals to the customers to have them reused manifold. This model offers clothing items as a rental service rather than for purchase. This model can extend the life of clothing items and reduce the overall consumption of new goods. It shifts the focus from ownership to access, allowing consumers to use products for a specific period. The PaaS model can contribute to a significant reduction in the demand for raw materials and waste generation. By ensuring that products are used for their entire lifespan, fashion companies can create value while maintaining control over their products, facilitating easier recycling or upcycling.

- Design for Longevity
  Designing products for longevity means creating clothing that is durable, repairable, and timeless. This model is crucial in the circular fashion ecosystem as it has essential qualities to increase its usage for a longer period. Niinimäki et al. (2020) emphasise that design for longevity is a foundational aspect of circular fashion. By prioritising quality and durability, companies can encourage consumers to buy less frequently, which in turn helps in reducing the environmental concerns that arise because of fashion product production and usage.
- Take-Back Schemes and Resale
  Take-back schemes involve companies taking the job of collecting the products used by their customers once the products reach their end of life. They ensure that they are either recycled, upcycled, or resold. Resale platforms enable customers to buy pre-owned clothing, which keeps garments in use and reduces the demand for new items.
- Clothing Swap
  This is another model that is helping in the growth of circular fashion. In this model, the clothes are swapped by the people through online and offline meets. Interested people come to a commonplace with their dress materials and exchange them with someone else's dress materials which are of their liking (Rathinamoorthy et al., 2017). People purchase clothing and after some time, their interest in wearing them is reduced and they look for some other new design. In such cases, they can attend such parties or meets to exchange their clothing with the ones' of their choice. Various organisations conduct such meetings, which help to avoid wastage of the textile materials and make them be in use for an extended period.

These models are not only helping in managing waste, but they also lead to the creation of new entrepreneurial ventures by utilising the market requirement for sustainable and pre-owned fashion. These models are very important in building circular fashion as the produced materials are being used in some way before ending up in landfills.

3 Growth of Business Models in Circular Fashion

Circular fashion models' growth is because of the shifting mindset in the fashion industry towards sustainability. Such models help in waste reduction and extension of the life of products with minimised environmental impact. The growth of circular fashion is being fuelled by multiple factors, including consumer demand, regulatory changes, and advancements in technology.

1. Increasing Consumer Demand for Sustainability

   Consumer awareness of environmental issues is on the rise, and this is directly influencing the growth of circular fashion models. The consumers are no more naïve as they have a lot of understanding towards the usage of sustainable products and are very conscious of buying products from brands that practice similar values concerning ethics, sustainability, etc. So the customers look for brands that do sustainable businesses and then buy from them. This mindset change in consumer preferences has led to a growing market for circular fashion models. The various models discussed in the above section like rental, swap, and resale are being employed. A study by Niinimäki et al. (2020) highlights that the important element in circular fashion is consumer demand. The study stated that the customers do not worry about the increased cost of the product due to sustainable manufacturing and are ready to spend the extra amount to purchase them, which encourages brands to adopt circular models. The rise of platforms like ThredUp and Rent the Runway illustrates this trend, as they cater to a market that values sustainability and circularity.

2. Regulatory Pressures and Government Initiatives

   Government regulations and policies also play an important role in the growth of circular fashion. Many governments are implementing regulations aimed at reducing waste and encouraging sustainable practices in the fashion industry. For instance, the circular economy action plan of the European Union includes measures specifically targeting the textile sector. They recommend the usage of materials that are recycled and also suggest working on improving product design for durability and recyclability. Kirchherr et al. (2018) discuss how these regulatory frameworks are accelerating the adoption of circular business models (CBMs). They note that companies are being incentivised to innovate and shift towards circular practices to comply with these regulations. Such measures help in decreasing the impact on the environment caused by the fashion industry. The study highlights that regulatory pressures are not only encouraging compliance but also driving competitive differentiation in the market.

3. Technological Advancements Facilitating Circular Models

   Technology is a key enabler of circular fashion models. Digital platforms are making it easier for consumers to participate in circular activities like renting, reselling, and recycling clothing. Moreover, technological innovations in material science are leading to the development of sustainable fabrics and recycling processes that are crucial for circular fashion. Bocken et al. (2016) explore the role of technology in supporting circular business models, noting that digital platforms enable the efficient exchange of pre-owned goods and support the logistics of rental and resale services. Additionally, advancements in recycling technology are allowing companies to recover and reuse materials more effectively, thereby closing the loop in fashion production.

4. Economic Opportunities and Competitive Advantages

   Circular business models offer significant economic opportunities for fashion companies. By adopting these models, companies can reduce costs associated with raw material procurement and waste management, create new revenue streams, and strengthen customer loyalty by aligning with consumer values. In their research, Bocken et al. (2019) mention that these models are also economically viable other than being environmentally friendly and sustainable. It is also conveyed in the study that clear product differentiation can be achieved by inculcating the circular fashion concept. This demarcation would lead to quality and environmentally conscious customers purchasing the products thereby ensuring constant business growth. This economic potential is a powerful incentive for the growth of circular fashion.

4 New Business Models: Rental, Resale, and Subscription Services

4.1 Reconceptualise Ownership: Rental, Resale, and Subscription Services in Fashion

The fashion industry is undergoing a significant shift. Consumers are now more aware of and updated with the environmental impact of fast fashion and are interested in more sustainable ways to engage with clothing. In response, innovative business models are emerging that challenge traditional notions of ownership. This chapter explores the rise of rental, resale, and subscription services in the fashion landscape.

4.1.1 A Shift in Consumer Values: Access Over Ownership

The concept of 'access over ownership' is gaining traction, particularly among younger generations. Consumers are prioritising experiences over material possessions and are drawn to models that allow them to experiment with different styles without the commitment of buying (Yuan & Shen, 2019).

- The Sharing Economy: Rental and subscription services capitalise on the sharing economy trend, where access to products takes precedence over ownership.
- Environmental Concerns: Rental, resale, and subscription services offer a more sustainable alternative by increasing the life of the apparel leading to a decrease in wastage and environmental load.
- Job Creation: These models often require new business structures, leading to the creation of jobs in logistics, customer service, technology, and maintenance.
- Increased Consumer Spending: While initial upfront costs are lower, consumers might spend more over time due to the convenience and flexibility offered by these services.
- Economic Growth: The rise of these services can contribute to overall economic growth by fostering innovation and creating new market opportunities.

- Resource Efficiency: By promoting sharing and reuse, these models can reduce waste and resource consumption, leading to potential cost savings for businesses and consumers.
- Stimulating Innovation: The need to manage inventory, optimise pricing, and provide excellent customer service can drive innovation in technology and business models.

4.1.2 Rental Services: Borrowing the Look

Clothing rental services allow users to borrow designer pieces for specific occasions or limited periods. This model caters to those seeking trendy styles without the high price tag or environmental burden of purchasing new clothes. Also, fashion renting provides an option to reuse clothing products and fulfil consumers' fashion needs (Chu, 2010).

- Special Occasion Wear: Renting a designer gown for a special event offers a more affordable and sustainable option compared to buying a one-time-wear piece.
- Trend Experimentation: Rental services allow users to experiment with high-end labels and trendy styles without the commitment of a full purchase.
- Challenges: Ensuring garment quality, dry cleaning costs, and limited size availability can be potential drawbacks for some users.

4.1.3 Resale Platforms: Repurposing Pre-owned Clothes

Resale platforms offer a space for consumers to sell and buy pre-owned clothing. This model promotes sustainability by increasing the life cycle of garments and diverting them from landfills (Arrigo, 2021).

- Reduced Environmental Impact: Resale platforms avoid the time and effort required in the manufacture of new products as the products manufactured already are resold, which leads to a reduction in the depletion of the environment.
- Treasure Hunting: Resale platforms offer the thrill of finding vintage pieces or gently used branded items at a fraction of the original cost.
- Challenges: Authenticity verification, quality control, and ensuring a smooth buying and selling experience are crucial considerations for resale platforms (Lang et al., 2020).

4.1.4 Subscription Services: A Curated Closet Delivered

Subscription services offer a curated selection of clothing delivered to your doorstep regularly. This model caters to those who desire a hassle-free way to refresh their wardrobe and explore new styles (Chen et al., 2018).

- Convenience and Personalisation: Subscription services curate boxes based on individual style preferences, eliminating the need for extensive shopping trips.
- Discovery and Experimentation: Subscription boxes allow users to experiment with new trends and receive professional styling advice.
- Challenges: The 'fit' guarantee can be tricky, and users may not always connect with every item they receive.

4.1.5 The Future of Fashion: A Blend of Models

The future of fashion is likely to be a blend of these new business models alongside traditional retail. This helps the customers with more choices to choose the mode that fits their preferences and requirements.

- Omnichannel Experience: Retailers may integrate rental, resale, and subscription elements into their existing business models, creating a seamless omnichannel experience for consumers.
- Focus on Quality and Sustainability: As the focus shifts towards access over ownership, quality and sustainability will become even more critical for fashion brands.
- Building a Community: These new models have the potential to foster a sense of community among users who share an interest in sustainable fashion and conscious consumption.

In conclusion, rental, resale, and subscription services are changing the fashion industry by offering consumers more sustainable ways to engage with clothing. These models present exciting opportunities for businesses to cater to a shifting consumer landscape and create a more circular fashion ecosystem that benefits both the environment and our wardrobes (Tao & Xu, 2018).

4.2 Challenges and Opportunities in Implementing Circular Business Models

The circular economy produces a compelling way to the traditional approach of 'take, make, dispose of', which was a linear model in practice. The circular business model goal is to ensure that the products are used for increased time till it is possible to be kept leading to reduced waste and impact on the environment. Implementing these models comes with its own set of challenges and opportunities. This chapter also discusses equipping businesses with the knowledge to navigate the circular economy effectively (Niinimaki et al., 2018).

4.2.1 Challenges: Hurdles on the Path to Circularity

Transitioning to a circular business model requires overcoming several hurdles. Here are some key challenges to consider:

- Shifting Mindsets: A significant challenge lies in changing mindsets, both within businesses and among consumers. Moving from the focus on selling products to offering services and prioritising product life extension requires a cultural shift.
- Design for Circularity: Products need to be designed with circularity in mind, considering factors like disassembly, remanufacturing, and recyclability. This often requires a rethinking of traditional design principles.
- Reverse Logistics: Establishing efficient reverse logistics practices for the collection of the products that are used is vital for CBMs. This can be complex and requires collaboration with various stakeholders.
- Consumer Behaviour: Encouraging consumers to embrace product sharing, repair, and reuse can be challenging. Educational initiatives and incentives are needed to promote a more circular mindset among consumers.
- Lack of Infrastructure: The infrastructure for material recovery and recycling may not be sufficiently developed in all regions, hindering the implementation of CBMs that rely on closed-loop systems.

4.2.2 Opportunities: Embracing the Circular Advantage

Despite the challenges, CBMs offer a multitude of opportunities for businesses. Here are some key benefits to consider:

- Resource Efficiency: By extending product lifespans and prioritising material reuse, CBMs can contribute to a greater reduction in utilisation of the company's resource consumption and environmental footprint.
- Cost Reduction: The process of reducing waste and increasing the effective utilisation of resources would lead to a reduction in cost and increased savings for businesses over the long term.
- Innovation and Differentiation: CBMs can foster innovation in product design, material science, and service offerings, differentiating businesses from competitors.
- Enhanced Brand Image: Consumers are increasingly drawn to sustainable brands. Implementing CBMs can strengthen a company's brand image and attract eco-based customers.
- Building Resilience: CBMs can help businesses become more resilient to fluctuations in raw material prices and resource availability.

4.2.3 Overcoming Obstacles: Strategies for Successful Implementation

Several strategies can help businesses mitigate the challenges and capitalise on the opportunities presented by CBMs:

- Collaboration: Collaboration with industry partners, NGOs, and governments is crucial for developing and scaling circular initiatives.
- Consumer Engagement: Educating consumers about the benefits of CBMs and creating user-friendly take-back and repair programmes are essential for fostering consumer buy-in.
- Policy Advocacy: Businesses can advocate for policies that support the development of the circular economy, such as extended producer responsibility schemes and investments in recycling infrastructure.
- Innovation: Continuous innovation in product design, material science, and service offerings is the key to overcoming technical challenges and creating commercially viable CBMs.
- Metrics and Measurement: Developing clear metrics to measure the environmental and economic benefits of CBMs is essential for demonstrating their value proposition to stakeholders.

Overall, following a circular economy can be hard for businesses, but it also offers big benefits. By considering everything and making smart changes, businesses can become more competitive and more resilient and can also help in creating a better future for the planet. This new way of doing business can help us grow without harming the environment.

5 Consumer Behaviour and Awareness

5.1 Impact of Consumer Behaviour Towards Sustainable Fashion

5.1.1 Sustainability in Style: Consumer Mindsets Shape the Future of Fashion

Consumers Are Driving Sustainable Fashion: Environmental awareness is on the rise, and people are making choices that reflect this. They are thinking more about ethics and the environment when they buy clothes. This section explores how consumer attitudes are pushing the fashion industry to be more sustainable. From style to sustainability, fashion used to be just about looking good and finding a good deal. Now, people are more aware of the impact their clothes have on the planet and the people who make them. Here is what is driving this shift:

- Outrage over Environmental Damage: Documentaries and news stories have exposed the environmental destruction caused by fast fashion, making people angry and demanding change (Cline, 2013).
- Demanding Fair Labour Practices: Consumers are concerned about how clothes are made and want brands to be transparent about their factories and how they treat workers.
- Eco-Influencers on Social Media: Social media stars who promote sustainable fashion are influencing what people want to buy. They show people eco-friendly options and brands that care about social responsibility.

5.1.2 The Power of Choice: How Consumer Attitudes Impact the Industry

Consumer attitudes have a crucial impact on the fashion industry in several ways:

- Market Demand: The growing demand for sustainable apparel is pushing brands to adopt more sustainable practices. Companies are increasingly offering eco-friendly collections made from recycled materials or organic fibres.
- Transparency and Traceability: Consumers are demanding transparency from brands regarding their sourcing and manufacturing practices. This pressure is driving the adoption of traceability initiatives to ensure ethical and sustainable production.
- Shifting Business Models: The rise of rental, resale, and subscription services caters to consumers who prioritise access over ownership and aligns with a more sustainable approach to fashion consumption.

5.1.3 Challenges and Opportunities: Navigating the New Landscape

While the increasing consumer interest in sustainable fashion is more and also positive, there are challenges to address:

- Greenwashing: Some brands may participate in 'greenwashing', making wrong claims about their sustainability efforts. Consumers need to be critical and research brands before making purchases.
- Affordability: Sustainable clothing can sometimes be more expensive than fast fashion options. Initiatives to promote transparency in pricing and educate consumers about the true cost of garments can help bridge this gap.
- Lack of Awareness: Not all consumers are completely aware of the environmental and social impacts of the fashion industry. Educational campaigns and initiatives are crucial to raise awareness and encourage sustainable consumption.

5.1.4 The Future of Fashion: Collaboration for a Sustainable Future

The future of fashion depends on a collaborative effort between brands, consumers, and policymakers. Here is what we can expect:

- Increased Consumer Power: Consumers will continue to hold brands accountable for their environmental impact and social impact. Brands that prioritise sustainability will gain a competitive advantage.
- Innovation in Materials and Production: Innovation in material science and production processes will lead to the development of more sustainable and eco-friendly textiles and garment construction methods.
- Policy and Regulation: Governments may implement policies that encourage sustainable practices and confine companies accountable for their environmental footprint.

Overall, consumers are becoming more knowledgeable of how their shopping habits affect the environment. By learning and making smart choices and by supporting sustainable brands consumers can make big differences. If everyone works together—shoppers, brands, and government—it would make fashion more sustainable and ensure that looking good does not harm the planet.

5.2 Strategies for Increasing Consumer Awareness and Engagement

5.2.1 Grip and Grow: Strategies for Expanding Customer Reach and Engagement

In today's crowded marketplace, capturing consumer attention and fostering engagement is critical for any brand's success. Gone are the days of interruptive advertising; consumers crave authenticity, transparency, and meaningful interactions. This section details effective strategies for increasing consumer awareness and building lasting engagement.

5.2.2 Understanding Your Audience: The Foundation for Success

The cornerstone of any notable engagement strategy lies in an intense understanding of your target audience. Here is where to begin:

- Customer Personas: Develop detailed customer personas that encapsulate your ideal customer's demographics, psychographics, and online behaviour. Understanding their needs and wants allows you to craft personalised communications and engagement approaches that resonate deeply.

- Social Listening: This is where brands need to listen to society to ascertain what they think about the brand and its products. This allows one to identify their interests, concerns, and preferred communication channels.
- Customer Feedback: The feedback can be obtained by floating surveys, polls using different media and also by customer reviews. The obtained feedback can be reviewed to understand the customer requirements and plan for the future.

5.2.3 Building Engagement: Fostering Two-Way Communication

Engagement goes beyond simply broadcasting content. Here is how to encourage active participation from your audience:

- Interactive Content: Include interactive elements like polls, quizzes, and Q&A sessions into your content strategy. This encourages participation and a sense of community.
- Social Media Contests and Giveaways: Contests can be conducted through social media and some prizes can be given for the winners to build more engagement.
- Community-Building: It is the process of building communities online and offline to facilitate more people to connect with each other to talk about the brand experiences and retail experiences.
- Quick Response Time: This is a major requirement. It is required that the retailer responds to the customer requirements, doubts, and enquiries in the shortest possible time politely and professionally and this will enhance the customer engagement to a great extent. In this fast-paced world, getting people to answer to doubts is seen as a boon and so when customers get answers to their queries, the brands and retailers are positioned better in the consumer's mind.
- Customer Support Channels: This is another area where a retailer can score over the competitors. Customers should be able to reach the retailer through multiple ways like chats, calls, letters, and e-mails for their queries or issues.
- Customer Feedback Integration: The customer will always feel happy and privileged when they are listened to and if their suggestions are getting incorporated, then they will be very loyal to the brand. Hence, it is required to integrate customer feedback with decision-making in the retail store and it can be in the areas of product promotion, visual merchandising, quality, etc.
- User-Generated Content: Customers would feel happy to share their brand experiences. So those stories must be captured and shared through social media, thereby enhancing brand reputation. These are contents that are given by the user and so it eases the promotional work of the brand or the retailer as this positive word-of-mouth contents will generate more customers for the brand.
- Co-Creation: Involve customers in the product development process or marketing campaigns.
- Influencer Partnerships: Influencers are influencing businesses to a great extent in the present context and so partnering with them will lead to better customer response.

- Customer Loyalty Programmes: Reward loyal customers with exclusive benefits and perks. It is required that such programmes are created with more thought process as customers always look for some kind of appreciation or rewards for their shopping activity.
- Event Planning: Host in-person or virtual events to bring customers together.

5.2.4 The Power of Authenticity: Transparency and Trust

Consumers crave authenticity. Here is how to build trust and transparency with your audience:

- Brand Storytelling: Share your brand story, values, and mission. Let your audience connect with the people behind the brand and understand what makes you unique.
- Transparency in Communication: Be open and honest in your communications. Address customer concerns promptly and proactively.
- Social Responsibility: Highlight your commitment to social and environmental responsibility. Consumers are drawn to brands that align with their values.
- Clear Product Information: Provide detailed descriptions of each item, including materials, condition, and care instructions.
- Pricing Clarity: Ensure transparent pricing, including any additional fees or deposits.
- Hygiene Standards: Communicate cleaning and sanitisation processes. Consider certifications or partnerships with cleaning services to bolster credibility.
- Return Policy: Outline a clear and fair return policy, building trust by demonstrating flexibility.
- Sustainability Focus: Highlight the environmental benefits of renting, such as reducing waste and conserving resources.
- Ethical Sourcing: It is required that the sourcing follows all ethical practices so that there will also be an emphasis on using materials that are sustainable and also lead to following fair practices in labour engagement for sourcing.
- Community Involvement: Demonstrate a commitment to the community through partnerships or charitable initiatives.
- Customer Support: Provide excellent customer service with prompt and helpful responses to inquiries.
- Data Privacy: Ensure the protection of customer data and transparent data usage policies.
- Quality Assurance: Invest in quality control measures to maintain the condition of rental items.
- Partnerships: Collaborate with reputable brands and designers to enhance credibility.

6 Policy and Regulatory Frameworks

6.1 *Government Policies and Regulations Promoting Circular Fashion*

6.1.1 Stitching Together a Sustainable Future: Government Policies and Regulations for Circular Fashion

The fashion industry, while a significant contributor to global economies, also carries a hefty environmental burden. Linear production models, characterised by 'take-make-dispose' practices, contribute to resource depletion, textile waste, and pollution. Circular fashion, with its emphasis on extending product life cycles and minimising waste, offers a promising alternative. This section includes the role of government policies and regulations in promoting a circular fashion ecosystem.

6.1.2 The Need for Intervention: Market Failures and Circularity

Market forces alone may not be sufficient to drive a widespread transition to circular fashion. Here is why government intervention is necessary:

- Externalities: The environmental cost and social costs of conventional fashion practices are not fully borne by producers. Government intervention can help internalise these externalities through policies like extended producer responsibility (EPR) schemes.
- Information Asymmetry: Consumers may lack comprehensive information about the environmental impact of clothing or the sustainability practices of brands. Government regulation can promote transparency and empower consumers to make informed choices.
- Coordination Failures: A successful circular fashion system needs collaboration across the entire supply chain starting from the conceptualisation to production followed by usage and end-of-life management. Government policies can incentivise collaboration and remove barriers to circular practices.

6.1.3 Policy Instruments: A Toolbox for Circular Change

Several policy instruments can be used to promote circular fashion:

- Extended Producer Responsibility (EPR): This policy makes the producer more responsible as they have to manage the products manufactured from cradle to grave. This ensures manufacturers are financially accountable for the collection and management of the materials manufactured by them at the end of their lifespan. Such practices lead to eco-friendly design principles as they will ensure minimal material usage in manufacturing to reduce cost.

- Green Procurement: Governments can leverage their purchasing power by prioritising sustainable clothing for public institutions, setting a positive example, and influencing the market.
- Eco-labelling and Transparency: Standardised eco-labels can help consumers identify sustainable clothing options. Regulation can promote transparency throughout the supply chain, allowing consumers to make choices.
- Taxation and Subsidies: Giving tax holidays or tax deferments for a specific period along with subsidies can incentivise companies to use sustainable measures like using recycled materials or developing innovative recycling technologies.
- Infrastructure Investment: Government investment in infrastructure for waste collection, sorting, and recycling of textiles is crucial for creating a viable circular system.

6.1.4 Challenges and Considerations: A Balanced Approach

Implementing effective policies for circular fashion requires addressing certain challenges:

- Harmonisation and Consistency: A patchwork of regulations across different regions can create confusion for businesses operating in global markets. Harmonisation efforts are crucial.
- Administrative Burden: Regulations should be designed to balance environmental goals with minimising administrative concerns on businesses, particularly small- and medium-sized enterprises (SMEs).
- Consumer Behaviour: Policies should be complemented by educational initiatives to raise consumer awareness and encourage sustainable consumption practices.

6.1.5 The Road Ahead: Building a Collaborative Future

The transition to a circular fashion ecosystem requires a combined effort between governments, businesses, and consumers.

- Public–Private Partnerships: Association between governments and the fashion industry can accelerate innovation, develop best practices, and create effective circular business models.
- Consumer Education: Raising consumer awareness about the environmental impact and social impact of fashion and educating them on caring for clothes and participating in circular initiatives like clothing swaps or rental services are crucial.
- Research and Development: Government funding for research and development in areas like material science and circular design principles is essential for long-term progress.

6.2 Global Initiatives and Collaborations for Sustainable Fashion

6.2.1 Weaving a Sustainable Web: Global Initiatives and Collaborations for Circular Fashion

The fashion industry faces a stark challenge to reconcile its love for creativity due to the environmental damage caused by conventional production methods. Thankfully, a wave of global initiatives and collaborations is rising to meet this challenge. This chapter explores how these efforts are showing a more sustainable future for fashion.

6.2.2 A Collective Awakening: The Need for Global Action

The environmental impact of the fashion industry is a global concern. Here is why international cooperation is the key:

- Global Supply Chains: The fashion industry operates globally, with raw materials sourced, garments manufactured, and products sold across continents. Sustainability efforts need to encompass the entire supply chain, requiring international collaboration.
- Harmonisation of Standards: A patchwork of environmental regulations across different countries can hinder progress. Global initiatives can promote the harmonisation of sustainability standards, creating a level-playing field for businesses.
- Shared Knowledge and Best Practices: Collaboration allows countries to learn from each other's successes and failures, accelerating progress towards a more sustainable fashion ecosystem.

6.2.3 A Symphony of Change: Key Players and Initiatives

Several global initiatives are driving the shift towards sustainable fashion:

- The UN Alliance for Sustainable Fashion: This multi-stakeholder platform brings together UN agencies, NGOs, and fashion industry leaders to promote sustainability across the fashion value chain.
- The Ellen MacArthur Foundation: This foundation's 'Make Fashion Circular' initiative works with leading brands, retailers, and policymakers to develop a global roadmap for a circular fashion system.
- The Sustainable Apparel Coalition (SAC): This industry-wide alliance works to improve the environmental impact and social impact of apparel and footwear products through tools like the Higg Index.

- Regional Collaborations: Regional initiatives like the European Union Circular Economy Package and the Asia Fashion Sustainability Summit address specific challenges and opportunities within their regions.

6.2.4 From Vision to Action: Examples of Collaboration

These initiatives translate into tangible action through various collaborations:

- Innovation Challenges: Global platforms bring together designers, scientists, and businesses to develop innovative materials, production processes, and circular business models.
- Transparency Initiatives: Collaborative efforts promote greater transparency throughout the supply chain, thereby giving customers more power to question brands for any delays or issues.
- Education and Capacity-Building: Global initiatives provide training and resources to educate stakeholders across the fashion industry on best practices for sustainable production and consumption.

6.2.5 Challenges and Opportunities: Building a United Front

Despite the progress, challenges remain:

- Stakeholder Engagement: Ensuring active participation from all stakeholders, including developing countries with limited resources, is crucial for achieving widespread impact.
- Monitoring and Enforcement: Effective monitoring and implementation mechanisms are needed to ensure compliance with sustainability standards.
- Consumer Awareness: Raising global consumer awareness about sustainable fashion and empowering them to make informed choices is essential for long-term success.

6.2.6 The Future of Fashion: Stitching a Sustainable Tapestry

The future of fashion lies in collaboration. By fostering international cooperation, sharing knowledge, and working towards harmonised standards, global initiatives can pave the way for a sustainable fashion ecosystem. This requires:

- Multi-stakeholder Partnerships: Continued collaboration between governments, NGOs, brands, and consumers is essential for driving systemic change.
- Investment in Innovation: Funding research and development of new materials, recycling technology, and circular business models is crucial for long-term progress.
- A Shift in Consumer Mindsets: Educating consumers and promoting conscious consumption habits are essential for sustainable fashion to make a mark.

7 The Economic Viability of Circular Fashion

Economic viability refers to an economy that operates profitably by following circular fashion principles such as minimising waste and maximising resource efficiency in the fashion industry. This involves ensuring that circular fashion practices are cost-effective, generate sufficient revenue, and align with consumer demand for sustainable products. Factors such as supply chain optimisation, innovation in materials and processes, regulatory compliance, and brand reputation contribute to the economic sustainability of circular fashion businesses. Adopting circular practices also fosters innovation, opening new revenue streams through rental services, resale platforms, and remanufacturing processes.

7.1 The Potential Economic Benefits of Circular Fashion for Retail Businesses and Retail Consumers

The circular fashion model offers a new approach to apparel retailing that benefits both businesses and consumers. Let us explore the potential economic advantages.

7.1.1 For Businesses

- Cost Savings: Circular fashion practices such as extending the lifetime of the product through redesigning materials and creating new ones lead to cost benefits. Extending the lifetime of the garments can be done by reusing, renting, and leasing and some brands even offer free repair services. This has benefits such as satisfying customers emotionally and reinforcing business value. For example, Patagonia 'Worn Wear' is a service offered by them, where the used and worn garments brought by the customer regardless of the brand are offered a free repair. This campaign mainly encourages people to take care of their garments and repair the necessary faults to ensure that they last as long as possible. This also educates people on how to care for their garments and maintain their functionality.
- Logistic optimisation: Package and postage costs play a main role in rental, particularly in mid-market and premium sectors; if the logistic cost remains low, the rental can be profitable.
- New Revenue Streams: Embracing circularity opens doors to innovative business models. Retailers can offer access-based models (renting, leasing, and sharing of garments), collection and resale, recycling, and reuse, which generate new revenue streams alongside traditional sales. Example, ThredUp: an online thrift store where customers can buy and sell second-hand garments.
- Le Tote: A subscription-based service that sends customers a curated selection of clothes and accessories each month.

- Enhanced Brand Image: Consumers are increasingly environmentally conscious. By adopting circular practices, businesses can carry their brand image as sustainable and responsible, attracting a loyal customer base. Brands like Patagonia and Eileen Fisher have strong customer bases due to their commitment to sustainability.
- Supply Chain Resilience: Circular approaches like closed-loop production (using recycled materials for new clothes) can make a business's supply chain less vulnerable to disruptions and shortages of raw materials. Circular fashion can streamline supply chains by reducing the need for new materials and focusing on efficient reuse and recycling processes.
- Resource Efficiency: Circular fashion helps retailers by optimising the usage of the resources and improving the utilisation of the resources which leads to reduced cost and energy.
- Innovation and Differentiation: By embracing the circular models, the brands can stand out uniquely and differentiate brands in a competitive industry.
- Customer Loyalty: Sustainable practice can attract and retain customers who are increasingly concerned about environmental impacts.
- Risk Mitigation: Having circular practices reduces the risk concerning supply chain and resource availability. Any disruptions in the supply chain and scarcity of resources can be effectively mitigated with proper planning.

7.1.2 For Consumers

- Affordability: Circular models like thrift stores or rentals provide quality garments at lower price points.
- Affordable Fashion: Consumers can purchase high-quality, pre-owned clothing at half of the price or even lower, making fashion more accessible and affordable.
- Access to High-Quality, Durable Products: Products designed for longevity offer better value over time, reducing the need for frequent replacements.
- Variety and Flexibility Through Rental Services: Renting clothing for special occasions or regular use can be more economical than buying new items, especially for high-end or rarely worn pieces.
- Sustainable Choices: Circular fashion empowers consumers to make environmentally friendly choices. They can extend the life of existing clothes and reduce their contribution to textile waste.
- Ethical Satisfaction: Choosing circular fashion supports ethical practices, providing consumers with the satisfaction of knowing that their purchases contribute to fair labour practices and reduced waste.
- Environmental Impact Reduction: Consumers can lower their environmental footprint by purchasing from brands that prioritise sustainability and use eco-friendly materials.
- Extended Product Lifespan: Circular fashion expands rather than increases the lifespan of garments and reduces the frequency and cost of new purchases.

7.2 The Impact of Technology on Circular Fashion

Technology is a game-changer in the world of circular fashion for apparel retailing.

7.2.1 Enabling Circular Practices

- Traceability: Blockchain technology can lead to easy tracing of material in the supply chain and thereby contribute to increased traceability of products leading to less wastage.
- Material Innovation: Biotechnology advancements are leading to the development of new methods to recycle and regenerate textile waste while maintaining fibre quality.
- Virtual Try-On: It is an augmented reality output that customers can try on virtually as per the name. This minimises waste generation which was caused by the need for having real garments to try and then return them if not found suitable in the case of online shopping.
- Enhanced Resale: Mobile apps can track a garment's life cycle and facilitate easy resale between consumers. Digital platforms can also connect consumers with repair services, extending the lifespan of clothing.
- Chemical Recycling: Technologies that break down the used and waste garments for reuse are highlighted as essential for circular fashion.

7.2.2 Empowering Consumers

- Informed Decisions: Digital tools provide information about the environmental impact of clothing choices and their benefits.
- Data-Driven Personalisation: Technology can personalise the shopping experience for circular fashion. For example, a 'digital wardrobe' app could suggest repair options or recommend resale platforms based on a user's clothing.
- Radio Frequency Identification (RFID) Technology: It enhances inventory management and helps customers find specific items in store or online.
- Omnichannel Retailing: This refers to utilising multiple ways of doing business with the customer including having physical and online stores to increase the shopping experiences and easiness for the customer as all the channels are integrated. It leads to seamless transitions between online and offline stores.
- Loyalty Programmes: Digital loyalty programmes reward customers with personalised offers and discounts based on their shopping behaviour.

8 Supply Chain Management and Transparency

Supply chain management and transparency are fundamental pillars of circular fashion in apparel retail. Supply chain management in circular fashion refers to the efficient and effective management of the flow of services, goods, resources, and information throughout the entire life cycle of fashion products, from raw material sourcing to production, distribution, use, and end-of-life disposal or recycling.

8.1 Building Sustainable and Transparent Supply Chains

8.1.1 Understanding Sustainable Supply Chains

A sustainable supply chain integrates ethical and environmentally responsible practices in a circular fashion model. According to Seuring and Müller (2008), sustainable supply chain management is concerned with achieving triple bottom-line outcomes—profit, planet, and people.

Key Concepts

- Environmental Impact: Minimising carbon footprints, reducing waste, and optimising resource efficiency.
- Social Responsibility: Social responsibility involves protecting labour rights, fair wages, and safe working conditions.
- Economic Viability: Maintaining profitability while investing in sustainable practices and technologies.
- Triple Bottom-Line: Balancing economic, environmental, and social performance in supply chain decisions.
- Life Cycle Assessment (LCA): Evaluating the environmental impact of products throughout their life cycle.

8.1.2 Transparency

Transparency is crucial for building trust and accountability within supply chains. It involves openly sharing information about processes, practices, and performance metrics across all stakeholders—from suppliers to consumers. As highlighted by Carter and Rogers (2008), transparent supply chains enhance visibility, mitigate risks, and foster collaboration towards sustainable goals.

- Benefits of Transparency.

  Risk Mitigation: Identifying and addressing potential risks such as supply disruptions, ethical violations, or environmental impacts.

Stakeholder Engagement: Building trust among consumers, investors, and regulators by demonstrating commitment to ethical and sustainable practices.

Operational Efficiency: Streamlining operations through improved data visibility and performance monitoring.

8.1.3 Challenges and Barriers

Despite the clear benefits, implementing sustainable and transparent supply chains poses significant challenges:

- Complexity and Fragmentation: Globalised supply chains often span multiple countries, cultures, and regulatory environments, complicating efforts to standardise practices.
- Cost Considerations: Initial investments in sustainable technologies or supplier audits may increase operational costs.
- Resistance to Change: Cultural inertia within organisations or reluctance among suppliers to disclose information can hinder transparency efforts.
- Consumer Perception: Half of the population is still not aware of sustainable and circular fashion which affects the demand for circular fashion.
- Fast Fashion Culture: The fast fashion model prioritises speed and low cost, which conflicts with the principles of sustainability.
- Supply Chain Complexity: The complexity of global supply chains makes it difficult to ensure sustainability through every stage.
- Industry Resistance: There may be resistance from established players in the fashion industry who are reluctant to change existing, profitable business models.

8.1.4 Strategies for Success

Achieving sustainable and transparent supply chains requires a multifaceted approach:

- Collaboration: Partnerships with suppliers, industry peers, and others to share and innovate best practices and innovations.
- Technology Adoption: Sage of digital tools such as blockchain for traceability or AI for predictive analytics to enhance transparency and efficiency.
- Regulatory Compliance: Stay informed and compliant with evolving global regulations on environmental protection and labour standards.
- Consumer Education: Empower consumers with information about sustainable products and encourage responsible purchasing decisions.
- Data-Driven Decisions: Utilise data analytics to make informed decisions about inventory, marketing, and customer preferences.
- Community Engagement: Build strong connections with local communities and engage in social and environmental initiatives.

8.2 *Role of Technology in Improving Supply Chain Transparency*

In the dynamic landscape of modern supply chains, transparency has become a cornerstone of ethical business practices and consumer trust. Technology transforms traditional supply chain operations by enhancing visibility, traceability, and accountability across global networks. This section explores how various technological innovations drive improvements in supply chain transparency, drawing insights from scholarly literature and industry practices.

8.2.1 Technological Innovations Enhancing Transparency

- Blockchain Technology

  Blockchain technology reduces manual tasks and paperwork associated with traditional transparency methods by managing the data. It has improved tracking, which leads to better inventory management and lowers logistic costs. It reduces the counterfeit products from entering the supply chains, which saves the retailer from the consequences caused by it such as fraudulent returns and lost revenue.

- Internet of Things (IoT)

  IoT is a solution to challenges such as the constant pressure on the manufacturing sector to increase efficiency and productivity. IoT helps in optimising production processes and reducing downtime by adding sensors to the physical objects.

- Artificial Intelligence (AI) and Predictive Analytics

  AI-powered analytics are transforming supply chain management by leveraging big data to predict trends, optimise inventory levels, and enhance operational efficiency. Predictive analytics enable proactive risk management, supplier performance evaluation, and route optimisation, thereby improving transparency and responsiveness across the supply chain.

- Radio Frequency Identification (RFID)

  Used for tracking and managing inventory in real time, RFID tags provide detailed information about the product's origin, production, and location, enhancing transparency throughout the supply chain.

- Digital Platforms

  These platforms, including traceability apps and websites, provide consumers with information about the sustainability practices of brands and their supply chains.

8.2.2 Challenges and Considerations

While technology offers significant benefits for supply chain transparency, several challenges must be addressed:

- Data Integration and Interoperability: Ensuring seamless integration of systems and data sources across stakeholders.
- Cybersecurity.
- Skills and Talent: Building capabilities within organisations to effectively utilise and manage advanced technologies.

8.2.3 Future Directions and Recommendations

To leverage technology for enhancing supply chain transparency, organisations should consider the following strategies:

- Investment in Innovation: Continuously exploring and adopting emerging technologies that enhance visibility and traceability.
- Collaboration and Standards: Establishing industry standards and collaborating with technology providers, regulators, and partners to promote transparency.
- Education and Training: Investing in training programmes that improve the employee's efficiency.

9 Challenges and Future Directions

The fashion industry faces frequent challenges. This topic examines the circularity barriers such as consumer behaviour, technology limitations, and supply chain complexities.

9.1 Barriers to the Adoption of Circular Fashion

9.1.1 Understanding Circular Fashion

Circular fashion aims to disrupt the linear model of fashion consumption, where garments are produced, consumed, and discarded. Instead, it advocates for strategies that prioritise longevity, reuse, repair, and recycling of materials. Key principles include:

- Design for Durability: Creating garments that withstand multiple wears and washes.
- Material Recycling: Recovering and reintegrating materials into new products or processes.
- Closed-Loop Systems: Establishing systems that facilitate the continuous reuse of materials within the supply chain.

9.1.2 Barriers to Adoption

- Consumer Behaviour and Perception

  Consumer behaviour is a significant barrier to the adoption of circular fashion. Despite increasing awareness of environmental issues, many consumers prioritise affordability, fashion trends, and convenience over sustainability. Research by Fletcher and Grose (2012) emphasises that consumer perceptions of value, aesthetics, and status often overshadow considerations of sustainability and circularity. Moreover, unfamiliarity with circular fashion concepts and the perceived complexity of participating in recycling programmes deter consumer engagement.

- Supply Chain Complexity

  The fashion supply chain is inherently complex, involving multiple stakeholders across global networks. From raw material extraction to manufacturing, distribution, and retailing, each stage presents challenges for implementing circular practices. According to Niinimäki et al. (2020), supply chain fragmentation and the lack of predetermined routine recycling processes affect the seamless integration of circular principles. Manufacturers often face logistical and operational challenges in sourcing recycled materials and maintaining product quality and consistency.

- Technological and Infrastructure Limitations

  Effective implementation of circular fashion relies heavily on technological advancements and infrastructure capabilities. Current textile recycling technologies struggle with mixed materials, chemical contaminants, and the scalability required to handle large volumes of textile waste (Niinimäki et al., 2020). Limited infrastructure for collecting, sorting, and processing of textile waste further impedes the circularity of fashion products, making it economically unfeasible for many stakeholders.

- Regulatory and Policy Frameworks

  The absence of regulatory frameworks and policies tailored to promote circular fashion is another barrier. Policies governing waste management, recycling incentives, and standards for sustainable practices vary widely across regions, complicating efforts to establish uniform guidelines and incentives for circularity (Fletcher, 2013). Moreover, inconsistent enforcement and compliance monitoring undermine efforts to encourage industry-wide adoption of sustainable practices.

9.1.3 Overcoming Barriers: Strategies and Recommendations

- Consumer Education and Engagement.
- Collaboration Across the Supply Chain.
- Investment in Research and Innovation.
- Policy Advocacy and Regulatory Support.

9.2 Future Trends and Innovations in Sustainable Apparel Retail

The future of apparel retail is undergoing a profound transformation driven by the imperative for sustainability. As consumers and industry stakeholders alike prioritise environmental and social responsibility, the landscape of sustainable apparel retail is evolving with new trends and innovations. This topic explores emerging trends, technological advancements, and innovative practices shaping the future of sustainable apparel retail.

9.2.1 Emerging Trends in Sustainable Apparel Retail

- Circular Economy Principles
  The adoption of circular economy principles is central to the future of sustainable apparel retail. Brands are increasingly embracing practices that extend the life cycle of garments, such as repair, resale, and rental services. According to PwC (2021), circular business models are gaining traction as they reduce waste, conserve resources, and appeal to eco-conscious consumers who seek durable and responsibly sourced products.
- Transparency and Supply Chain Traceability
  Transparency in supply chains is becoming a non-negotiable requirement for sustainable apparel retail. Consumers demand visibility into sourcing practices, production processes, and the environmental footprint of products. The latest technologies help in achieving transparency and traceability
- Digitalisation and E-Commerce Innovations
  Digital technologies are revolutionising the way apparel is marketed, sold, and consumed. E-commerce platforms facilitate direct-to-consumer (DTC) models that reduce the need for traditional retail infrastructure and optimise inventory management. AI-powered personalisation algorithms tailor shopping experiences, recommend sustainable alternatives, and minimise returns by matching products to consumer preferences.

9.2.2 Technological Advancements Driving Innovation

- Blockchain for Transparency
  Blockchain technology holds immense promise for enhancing supply chain transparency in sustainable apparel retail. By providing an immutable record of transactions and product provenance, blockchain improves accountability and reduces the risk of greenwashing. This transparency helps consumers find their choice of purchase while the brands will be responsible for their sustainable claims (Tapscott & Tapscott, 2016).

- Advanced Materials and Textile Innovations
  Innovations in material science are paving the way for sustainable alternatives to conventional textiles. Bio-based fibres, recycled materials, and textiles derived from agricultural waste offer eco-friendly alternatives that reduce reliance on virgin resources and mitigate environmental impacts (Niinimäki et al., 2020). These innovations cater to consumer preferences for sustainable and ethically sourced products.

9.2.3 Future Directions and Strategic Imperatives

- Collaborative Supply Chain Partnerships
  Collaboration across the apparel supply chain is essential for advancing sustainability goals. Partnerships between brands, suppliers, recyclers, and technology providers facilitate innovation in recycling technologies, sustainable sourcing practices, and circular business models (Fletcher, 2013).
- Consumer Education and Engagement
  Awareness of the environmental and social impacts of fashion consumption on consumers is crucial for driving demand for sustainable apparel. Brands can leverage storytelling, transparent communication, and educational campaigns to help consumers make choices and support sustainable brands. Consumer engagement initiatives foster loyalty and advocacy among environmentally conscious shoppers (Fletcher, 2013).
- Policy Advocacy and Regulatory Support
  Governments and industry associations play a pivotal role in fostering a conducive regulatory environment for sustainable apparel retail. Policies that incentivise eco-friendly practices, support research and development of sustainable technologies, and promote transparency in supply chains are essential for scaling sustainable innovations.

9.3 Innovations and Initiatives Driving Progress

- Closed-Loops System and Circular Economy Initiatives
  Companies and organisations are increasingly adopting closed-loop systems and circular economy principles to promote textile recycling. Initiatives such as clothes collection services, rental and resale platforms, and brand collaborations with recycling facilities aim to decrease textile waste from landfills and reintroduce recycled materials into the supply chain.
- Advanced Sorting and AI Technologies
  Advanced sorting of textiles by using near-infrared (NIR) spectroscopy and artificial intelligence (AI) are enhancing the precision and efficiency of textile waste sorting. AI algorithms can identify and sort textiles based on fibre composition,

colour, and condition, enabling more effective separation of recyclable materials and reducing contamination.

9.4 Future Directions

- Scaling Up Recycling Infrastructure
  Scaling up recycling infrastructure and capabilities globally is essential for achieving widespread adoption of textile recycling. Investments in recycling facilities, technology development, and collaboration between stakeholders across the textile supply chain are needed to meet the growing demand for sustainable textile solutions.
- Policy Support and Industry Collaboration
  Policy frameworks that incentivise textile recycling and support research and development are crucial for accelerating innovation in this field. Collaboration between governments, industry associations, and academic institutions can drive regulatory reforms and foster a supportive environment for sustainable textile recycling practices.

Advancements in textile recycling technologies hold promise for transforming the textile and apparel industry towards greater sustainability. By addressing technological challenges, scaling up recycling infrastructure, and fostering collaboration across stakeholders, the industry can reduce its environmental footprint, conserve resources, and promote a circular economy. As innovations continue to evolve and awareness of textile waste grows, textile recycling is poised to play an important role in creating a more sustainable future for fashion.

10 Conclusion

Circular fashion is essential for transforming the apparel industry into a sustainable and resource-efficient system. By adopting key models such as closed-loop recycling, products-as-a-service, and take-back schemes, circular fashion significantly reduces waste, extends product life, and promotes resource efficiency. Despite challenges such as consumer behaviour, quality control in rental and resale models, and regulatory gaps, the growing demand for sustainability and advancements in technology provide substantial opportunities for progress. Strong government action, global collaboration, and innovative policies are vital for overcoming barriers and accelerating the transition to circular fashion. Technological advancements, including blockchain and digital tools, play a critical role in enhancing supply chain transparency and efficiency. As the fashion industry faces greater pressure to address environmental impacts, circular fashion represents a pivotal shift towards achieving long-term sustainability goals. Embracing circular fashion not only benefits businesses through cost savings and new revenue streams but also offers consumers sustainable and affordable fashion choices. Thus, circular fashion is essential for encouraging a sustainable future in the apparel industry.

References

Arrigo, E. (2021). Digital platforms in fashion rental: A business model analysis. *Journal of Fashion Marketing and Management: An International Journal, 26*(1), 1–20.

Chen, T., Fenyo, K., Yang, S., & Zhang, J. (2018). *Thinking inside the subscription box: New research on e-commerce consumers* (pp. 1–9). McKinsey & Company.

Chu, H. (2010). A conceptual model of consumer pricing for online resale. In *2010 2nd IEEE international conference on information management and engineering* (pp. 578–580). IEEE.

Fletcher, K. (2013). *Sustainable fashion and textiles: Design journeys*. Routledge.

Fletcher, K., & Grose, L. (2012). *Fashion & sustainability: Design for change*. Hachette.

Lang, C., Li, M., & Zhao, L. (2020). Understanding consumers' online fashion renting experiences: A text-mining approach. *Sustainable Production and Consumption, 21*, 132–144.

Niinimäki, K., Peters, G., Dahlbo, H., Perry, P., Rissanen, T., & Gwilt, A. (2020). The environmental price of fast fashion. *Nature Reviews Earth & Environment, 1*(4), 189–200.

Rathinamoorthy, R., Surjit, R., & Karthik, T. (2017). Clothing swap: Gateway to sustainable eco-friendly fashion. In *Handbook of ecomaterials* (pp. 1–24). Springer.

Seuring, S., & Müller, M. (2008). From a literature review to a conceptual framework for sustainable supply chain management. *Journal of Cleaner Production, 16*(15), 1699–1710.

Tao, Q., & Xu, Y. (2018). Fashion subscription retailing: An exploratory study of consumer perceptions. *Journal of Fashion Marketing and Management: An International Journal, 22*(4), 494–508.

Tapscott, D., & Tapscott, A. (2016). *Blockchain revolution: How the technology behind bitcoin is changing money, business, and the world*. Penguin.

Yuan, Q., & Shen, B. (2019). Renting fashion with strategic customers in the sharing economy. *International Journal of Production Economics, 218*, 185–195.

Circular Fashion for Sustainability: Materials, Design, and Technology

R. Surjit, C. Harani, R. Akshya, and R. Akshatha

1 Introduction

In a circular fashion system, the production and end of product are valued equally. The main aim is to reduce waste, which can be achieved by reusing, repairing, recycling, and biodegrading things. Though it is a vast and complex idea, circular fashion sees the fashion business as an almost never-ending cycle where materials are used for as long as feasible. This idea is around understanding the origins of objects, their materials, the makers, and our responsibility for their entire life cycle. It is said that circular fashion is a development from the circular economy and this term was coined in 2014 by two people at different places namely Dr. Brismar and a staff of a leading apparel brand H&M (Rathinamoorthy, 2019). Circular fashion involves the concept of utilising apparel and dress materials to the maximum possible extent by reusing, recycling or extending their usage by any means. It also involves processes like rentals, resale, and subscription services to reduce material wastage. The new generation techniques like clothing swap also help in reducing the material going as waste quickly. The circular fashion works in such a way that it extends the product's use than leading it to wastage and landfills, thereby protecting the environment and reducing the pollution.

The circular fashion involves various principles as listed by Bismar (2020). There are 16 of them. They state that the materials should have biodegradability, longevity, resource efficiency, recyclability, and purpose. One should source more local materials that are without toxicity, renewable, and are produced ethically and

R. Surjit (✉) · C. Harani · R. Akshya · R. Akshatha
Department of Fashion Technology, PSG College of Technology,
Coimbatore, Tamil Nadu, India
e-mail: rst.fashion@psgtech.ac.in; 22h207@psgtech.ac.in; 22h203@psgtech.ac.in;
22h202@psgtech.ac.in

S. S. Muthu (ed.), *Circular Business Models in the Apparel Industry*,
Environmental Footprints and Eco-design of Products and Processes,
https://doi.org/10.1007/978-3-031-80193-8_7

efficiently. They should support a long life or have a long life. The other principles talk about buying quality products that can be reused, recycled, or composted. It is also required that the materials should be easy to use, wash, and repair. There should be more collaboration to rent, loan, swap, or redesign the products so that the environment can be protected. The advent of fast fashion has led to a total disruption in the ecology of textiles and apparel as the industry is causing more pollution than ever leading to increased environmental concerns. Fast fashion involves synthetic materials that are not friendly to the environment and such materials take years to degrade. However such materials are used in abundance, thereby leading to environmental pollution. In the case of the usage of polyester fibre, a report by Gina-Marie (2016) states that 21.3 million tonnes of polyester was used in 2017 for manufacturing. It is an increase of more than 150% of what was being used in the 2000s. Such increased usage affects the environment and makes it difficult in the future to sustain. Hence, circular fashion is an alternative to this as the materials are being reused, recycled, rented, swapped, etc., to keep them in use for longer periods. It involves even using synthetic material products for a longer duration, thereby reducing wastage. There is a lot of support from the government and legal regulatory frameworks that support circular fashion. The consumer awareness is also helping circular fashion grow as consumers understand the importance of buying from environmentally conscious retailers.

The circular fashion would be an effective alternative to fast fashion as it involves utilising eco-friendly materials for production and greener manufacturing practices. The product designing is also done in such a way that the garments withstand the test of time and durable for longer periods. In this chapter, an attempt has been made to discuss the materials, methods, and designs that are used to build circular fashion.

1.1 Definition

An economic model known as a "circular economy"is founded on Indigenous principles and strives to maximise resource utilisation while minimising waste. It is a closed-loop material flow in the whole economic system. The various dimensions of it are reduction, reuse, recycling, economic prosperity, and environmental quality (Kirchherr et al., 2017). The circular economy is not a linear economy, which is produce–use–dispose. Circular fashion places a strong emphasis on closed-loop systems, longevity and reusability, and resource efficiency and regeneration. The goal of circular fashion is to rethink the design, production, use, and disposal of clothing to establish a closed-loop system. Implementing circular fashion leads to reduced pollution and waste, thereby reducing the need for more raw materials to produce an intended product as it advocates endless reuse and recycling. This change provides a remarkable solution to the environmental problems faced by the industry, opening doors for creative business concepts and a more ecologically aware consumer base. Circular fashion addresses the key issues faced by the fashion industry and also helps in areas like supporting repair services, encouraging people to mend or alter clothing, and facilitating recycling or upcycling.

1.2 Fashion Sustainability's Evolution and Historical Context

Fashion's embrace of sustainability is not a new idea. Let us examine a condensed timeline of this developing discussion:

- Early twentieth century (1900s): The emergence of synthetic fabrics, such as polyester and nylon, created new design opportunities but also sparked questions about their long-term sustainability and possible effects on the environment. Unlike natural fibres like cotton, these materials could build up in landfills because they are not easily biodegradable.
- 1960s and 1970s: The environmental movement became increasingly popular, bringing attention to the pollution that comes from industrial processes, such as the manufacture of textiles. Concerns about the use of dangerous chemicals in fabric finishes and dyes have surfaced, drawing attention to the possible health risks to both consumers and employees.
- 1990s and the beginning of the 2000s: Recycled materials and organic cotton became more popular as environmentally friendly substitutes. As a result of growing consumer awareness of the negative environmental and social effects of conventional fashion, more ethical and sustainable practices have taken hold. Companies started experimenting with the production of organic cotton and incorporating recycled materials into their limited-edition collections.

Beginning in the late 2000s, there was significant concern over the rise of fast fashion companies due to their constant release of new collections at low prices. The fast fashion industry's "disposable clothing" mentality, increased production volume, and reliance on synthetic materials have all contributed to the industry's exploding environmental footprint. During this time, the human cost of inexpensive apparel was brought to light by the growing awareness of labour practices throughout the global fashion supply chain (Carter & Rogers, 2008). The evolution of circular fashion in the present shows the effects of the fashion industry on the environment and society clearly, and a growing number of companies, designers, and customers have begun adhering to the principles of circular fashion. An increasingly sustainable future for fashion is being ushered in by technological developments in textile recycling and creative business strategies like rental services. Circular fashion plays a significant role in reshaping the fashion industry due to growing consumer demand for transparency and environmentally friendly options. To achieve a "sustainable" industry, there is a constant need for innovation and collaboration in the ongoing journey towards circular fashion. The foundation for comprehending its fundamental ideas and historical setting has been established in this chapter.

2 Fashion's Environmental Impact

The fashion industry, while significant for cultural expression and economic activity, also has a considerable negative impact on the environment. These various environmental impacts are discussed below:

- Water Consumption: Textile production is a highly water-intensive process. Cotton requires a large amount of water for cultivation. It is estimated that a cotton t-shirt can consume thousands of litres of water, which is equivalent to a human's drinking needs for several years. Synthetic fibres, while appearing less water-intensive initially, still necessitate significant water usage during processing and energy generation.
- Chemical Contamination: Dyeing and finishing textiles involve a complex mix of chemicals, many of which are harmful. Improper disposal of wastewater from these processes can contaminate waterways, impacting aquatic ecosystems and potentially entering the human water supply.
- Microplastic Threat: Synthetic clothing sheds microplastic fibres during washing. These tiny plastics pollute oceans and infiltrate the food chain, harming marine life and potentially accumulating in humans who consume seafood.
- Greenhouse Gas Emissions: The textile cycle, from extraction of raw materials to production, logistics, and disposal, contributes majorly to greenhouse gas emissions. Synthetic fibres often rely on fossil fuels in their production, further exacerbating climate change.
- Fast Fashion's Footprint: The "fast fashion" model, characterised by cheap, trendy clothing with short lifespans, encourages overconsumption and frequent disposal. This constant churn of clothing fuels the industry's environmental impact.
- Landfill Burden: A substantial amount of discarded clothing ends up in landfills, which releases harmful methane gas while decomposing.
- Recycling Challenges: Current textile recycling technology struggles to efficiently convert used clothing back into high-quality fibres. This limitation creates a dependence on raw materials and hinders progress towards a more sustainable industry. The rise of organic cotton production, closed-loop recycling systems, and innovative bio-based fibres offers hope for a greener future.
- Consumer Responsibility: Consumers play a vital role in making informed choices. Buying less clothing, choosing high-quality garments that last longer, opting for sustainable brands, and embracing second-hand clothing options are all impactful actions.

- Sustainable fashion is an approach towards the creation of products that prioritise environmental and social concerns (Priyadarshini & Prasad, 2024).

3 Principles of Circular Economy in the Fashion Industry

The circular economy works in the concept of "Reduce, Reuse, Recycle, and Redesign" where one sees how to completely remove waste and avoid pollution, upcycle the products, and conserve nature by regeneration.

3.1 Reduce

The first step is reducing our overall clothing consumption. This starts with a shift in mindset. Instead of chasing fleeting trends, customers should buy high-quality garments produced from durable materials such as organic cotton, linen, or recycled synthetics. These pieces will withstand the time, reducing the need for replacements. Focusing on timeless styles that transcend fleeting trends is the need of the hour. A classic trench coat or a well-tailored blazer will remain wardrobe staples for years. Furthermore, promoting responsible consumerism is the key. Encouraging mending small tears, replacing buttons, or learning basic sewing skills to extend the life of existing pieces will help in the reduction of garments thrown away as waste.

3.2 Reuse

Extending the lifespan of clothing goes beyond personal care. Embrace the vibrant world of pre-owned clothing. Platforms like ThredUP, Poshmark, and vintage stores offer a treasure trove of unique finds and a sustainable alternative to new purchases. For occasional wear items like formal dresses or seasonal pieces, consider renting instead of buying. This reduces overall production and allows for a wider variety in one's wardrobe without the burden of ownership. Finally, do not underestimate the power of upcycling. Transform unwanted clothes into something entirely new, like a tote bag from old jeans, or use scraps for crafting projects.

3.3 Recycle

When garments finally reach the end of their usable life, recycling technologies offer a solution. Polyester, nylon, and even cotton can be reborn as new clothing through pre-consumer or post-consumer recycled materials. This reduces reliance on raw resources and decreases environmental impact. Exciting advancements in mechanical and chemical recycling are improving the efficiency and quality of recycled textiles, allowing for closed-loop systems where old clothes are transformed back into new ones. Additionally, natural fibres like hemp and organic cotton offer a biodegradable alternative, minimising landfill waste at the garment's final stage. Textile waste is a major environmental concern. Recycling technologies offer a solution by giving new life to discarded clothing and textiles. Here are some key innovations tackling this challenge:

- Enhanced Mechanical Recycling: The traditional method of mechanically breaking down used textiles has limitations. Fibres can become shorter and weaker with each recycling process. However, advancements are addressing this issue. New technologies like friction spinning can handle a wider range of textile waste, including blends and lower-quality materials. Additionally, innovations in fibre blending and re-engineering allow for the creation of high-quality recycled fabrics with properties closer to virgin materials.
- Chemical Recycling: This exciting technology breaks down used textiles on a molecular level, essentially depolymerising them back into their base chemicals. These chemicals can then be used to create virgin-quality recycled materials. Chemical recycling offers significant potential for closing the loop in textile production. It allows for the recycling of a wider variety of materials, including blended fabrics and even heavily soiled textiles that are often challenging for mechanical recycling.

Despite the significant advancements, challenges remain in textile recycling. The sorting and separation of different fibre types are crucial for efficient recycling, but this can be a complex and labour-intensive process. Additionally, the scalability and cost-effectiveness of some recycling technologies are still evolving.

3.4 Redesign

Circular principles should be embedded in the design process itself. Garments should be easy to take apart, facilitating material separation and efficient recycling later. Modular design with interchangeable components allows for easy updates and personalisation, extending the lifespan and adaptability of a piece. Furthermore, using safe dyes and finishes that do not hinder future recycling processes or release harmful chemicals is crucial for a truly sustainable system (Fransen, 2023).

4 Overview of Sustainable and Eco-Friendly Materials

One of the main elements of sustainable fashion is the use of sustainable products that do not affect nature much, such as natural fibres over artificial fibres rather than synthetic fabrics. It also involves educating the customers on the importance of choosing quality over quantity and regulating the consumption pattern of the market. As regards production, ethical production, which is the first step to begin with, must initially involve the proper treatment of working staff providing a cordial environment and following practices such as reducing water usage and maintaining safety standards. The shift to sustainable fashion is inclusive of the use of renewable energies for production and reusable and easily degradable materials for manufacturing. The following are some of the widely preferred materials for sustainable materials:

- Recycling Cellulosic Fibre: Recycling cellulosic fibres from textile waste is a promising approach to reduce environmental impact. These fibres can be recovered using chemicals like NaOH, phosphoric acid, or methylmorpholine-N-oxide. Additionally, bacteria can convert cellulosic waste into bacterial cellulose, a valuable material for the biomedical industry. This process is more sustainable than using traditional feedstocks like sucrose or glucose, which are expensive and produce lower yields. By utilising textile waste, we can conserve resources and decrease production costs. Essentially, recycling cellulosic fibres offers a greener alternative to producing new materials from scratch.
- Recycling Polyester: Polyester is a dominant and durable synthetic fibre with growing demand. Its production is inexpensive, and it resists damage from chemicals and heat. However, polyester takes centuries to degrade, making recycling essential for environmental protection.
- Recycling polyester involves three primary methods:

    Thermal Recycling: Melting polyester waste to create new fibres is a relatively simple but less effective process over time.
    Chemical Recycling: Breaking down polyester into its original components is costly and requires strict safety measures.
    Biological recycling: Using enzymes to degrade polyester, offering a more environmentally friendly approach but still under development.

- Wool Reuse: Wool is prone to shrinkage and felting, making it difficult to use in its raw form. To overcome this, harsh chemicals are typically used to treat the fibres, posing significant environmental risks. Alternative treatments such as enzymatic and plasma processing are being explored to reduce chemical use. However, these methods have limitations, including slow processing time, potential fibre damage, and the generation of waste products. Wool offers potential as a sustainable material. Waste wool fibres can be repurposed as a construction material, contributing to a more sustainable future.
- Organic Cotton: Organic cotton is a sustainable alternative to the traditional cotton fibres produced. These are produced without the usage of harmful chemicals and they offer environmental, health, and economic benefits. By eliminating pesticides, herbicides, and synthetic fertilizers, organic cotton reduces pollution, improves soil health, and creates a safer working environment for farmers. The resulting fabric is often softer and hypoallergenic. While organic cotton cultivation is more complex and yields less per acre than conventional methods, its positive impact on people and the planet makes it a valuable choice for the textile industry (Nayak, 2022).

- Organic Wool: Organic wool is produced through humane and environmentally friendly methods. Sheep are raised without chemicals, hormones, or harmful pesticides, and their grazing land is free from synthetic treatments. While organic wool shares similar fibre properties with conventional wool, it prioritises animal welfare and environmental sustainability. To be certified as organic, wool must meet strict standards, including organic feed, no genetic engineering, and minimal chemical treatments. The growing demand for organic wool reflects a global shift towards more sustainable and ethical textile production.
- Wild Silk: Wild silk is produced in a sustainable and eco-friendly manner. Unlike cultivated silk, wild silkworms thrive in forests without human intervention or harmful chemicals. This process preserves forests and provides income for local communities. Wild silk is considered "vegetarian" because the moths are allowed to emerge from their cocoons before harvesting. However, the resulting silk fibres are shorter and less refined compared to cultivated silk. This means wild silk fabrics have a unique texture and natural sheen but require different processing techniques.
- Hemp: Hemp is a highly sustainable and durable natural fibre with a long history of use. It grows rapidly, requires minimal water, and can be cultivated organically without relying on harmful chemicals or genetically modified seeds. Hemp fabrics are strong, breathable, and resistant to mould and mildew. While hemp offers many environmental benefits, its production can have some drawbacks. The retting process, used to separate fibres from the plant, can pollute water sources. However, alternative methods like dew retting, steam explosion, and enzymatic retting are being developed to address this issue. Hemp cultivation remains illegal in many countries due to its close relation to marijuana, hindering its widespread use (Bell, 2023).
- Flax: Flax is a highly sustainable textile fibre. Compared to cotton, flax requires fewer resources, including water, pesticides, and energy. Its cultivation benefits the environment by improving soil health, reducing erosion, and absorbing carbon dioxide. Flax fibres are strong, absorbent, and durable, making them a desirable alternative to synthetic fibres. Additionally, the production of flax fabrics consumes less energy than other materials like glass and carbon fibre. Even after its use as a textile, flax can be repurposed into composite materials for various industries.
- Bamboo: It is a rapidly growing, sustainable plant that offers several advantages as a textile fibre. It is strong, moisture-wicking, and naturally antimicrobial. Bamboo cultivation requires minimal water and no harmful chemicals, making it an environmentally friendly choice. The process of turning bamboo into fabric can vary significantly in terms of sustainability. Chemical methods, similar to those used for rayon production, involve the use of toxic chemicals and raise environmental concerns. On the other hand, mechanical processes that use enzymes to extract fibres are considered more eco-friendly. Bamboo fibre has potential as a sustainable textile material, especially when produced using environmentally friendly methods. Its versatility allows for blending with other fibres and its use in various applications beyond clothing, such as polymer composites.

- Polylactic Acid (PLA) Fibre: PLA fibre is a biodegradable alternative to traditional, petroleum-based synthetic fibres. These fibres are obtained from renewable resources. Corn starch is one of the sources of these fibres. It offers potential environmental benefits. It leads to reduced energy utilisation and also lower levels of greenhouse gas emissions. PLA's commercial viability is hindered by several challenges. The production process competes for valuable agricultural land and resources traditionally used for food production. Additionally, while PLA is biodegradable, its breakdown in landfills can produce harmful methane gas. Its higher cost compared to synthetic fibres and its susceptibility to degradation during processing further limit its widespread adoption.
- Lyocell: Lyocell is a sustainable, plant-based fibre derived from eucalyptus trees. Its production process is environmentally friendly, using minimal water and pesticides, while effectively recycling the solvents involved. Lyocell fabrics are known for their softness, breathability, and wrinkle resistance. It is a sustainable alternative to other rayon fibres, but lyocell does have a drawback. The wetting process can weaken the fibres and cause slight damage. Overall, lyocell offers a promising balance of environmental benefits and desirable fabric properties.
- Soybean Protein Fibre: It is a sustainable textile made from soybean hulls, a by-product of food production. It offers a soft, luxurious feel similar to cashmere, along with excellent moisture-wicking and antibacterial properties. While durable, it is not as strong as cotton or hemp. Despite its potential benefits, commercial production of soybean fibre has been limited. This fibre represents a promising alternative to traditional, less sustainable materials, but further development is needed to fully realise its potential.
- Casein Fibre: It is derived from milk protein and offers a unique blend of natural and synthetic fibre qualities. It has a silky appearance. It has good breathability and moisture-wicking properties. These properties make it a convenient fibre to use for areas where comfortability is required. The fibre's natural antibacterial qualities and skin-friendly pH contribute to its health benefits. As a biodegradable, renewable resource, casein fibre is considered an eco-friendly and sustainable textile option with a wide range of potential applications in the apparel industry.
- Lenpur Fibre: It is a sustainable, cellulose-based textile made from wood pulp. It offers several advantages over other cellulosic fibres, including exceptional softness, high moisture absorption, and natural odour resistance. Lenpur is versatile, blending well with other fibres and producing smooth yarns suitable for various textile applications. Its comfortable properties make it a potential replacement for cotton in many products, and it is even used in high-end fashion items.
- Sea Cell Fibre: It is a unique textile derived from seaweed. Known for its health benefits, it contains essential nutrients like iodine and vitamins that can nourish the skin. The fibre's antimicrobial and anti-inflammatory properties contribute to its skin-friendly nature.
- Chitin and Chitosan Fibre: These are natural substances derived from sources like shellfish and fungi. These materials have been used for centuries and are now gaining attention for their potential in textiles. Both chitin and chitosan fibres are created through a chemical process and possess properties similar to

cellulose. They are known for their antibacterial and biodegradable qualities, making them suitable for applications in medical textiles and wound dressings. However, their production process and potential for commercialisation as apparel fibres are still under development.

- Coconut Shell Fibre: Coconut shells, often overlooked, are being transformed into a valuable resource for the fashion industry. This innovative fabric offers several advantages, including durability, strength, and moisture resistance. By utilising coconut shells, manufacturers can reduce reliance on synthetic materials, promoting sustainability. This eco-friendly fabric is biodegradable, recyclable, and has a positive environmental impact. Its versatility makes it suitable for various applications, with sportswear being a prominent example. Brands like North Face have incorporated coconut shell-based fabrics into their products, demonstrating the material's potential in high-performance apparel.
- Eucalyptus Yarn: It is a sustainable textile produced from trees grown in areas unsuitable for agriculture. This eco-friendly process reduces land use and minimises environmental impact. The resulting yarn is known for its softness, absorbency, and cooling properties, making it a desirable alternative to traditional fibres like cotton and silk. The production process involves converting eucalyptus wood into pulp, which is then spun into yarn. The manufacturing methods are designed to be environmentally friendly, with water and solvents being reused.
- MycoTEX: It is a groundbreaking, biodegradable fabric created from mycelium, the root structure of fungi. This innovative material offers a sustainable alternative to traditional textiles. Grown in a laboratory, MycoTEX requires minimal resources, including water, land, and chemicals. The fabric is customisable, adaptable, and completely compostable at the end of its life cycle. Its unique properties and eco-friendly production process have garnered significant attention from the fashion industry and consumers alike. MycoTEX represents a promising future for sustainable fashion (Raman et al., 2022).
- Pinatex: It is an innovative, vegan leather substitute created from pineapple leaf fibres. This eco-friendly material is produced from waste products of the pineapple industry, making it a sustainable and ethical choice. With properties similar to leather, such as strength, flexibility, and breathability, Pinatex can be used for a variety of products, including fashion accessories, footwear, and even automotive interiors. The production process is environmentally friendly, with minimal waste and a positive impact on the communities involved in pineapple cultivation. Pinatex represents a significant step forward in the fashion industry's pursuit of sustainable materials.
- Mushroom-Based Leather: It is a promising eco-friendly alternative to traditional leather. The root-like structure of mushrooms, called mycelium, can be transformed into a durable material through specific treatments. This research focused on creating and analysing mushroom leather, exploring its properties such as strength, flexibility, and resistance to environmental factors. By examining the material's structure and composition, researchers aim to develop a sustainable leather substitute with comparable or superior qualities to animal and synthetic leathers.

5 Methods of Manufacturing

It requires sustainable manufacturing practices to avoid problems to the environment and also reduce wastage during production and post-production. Hence, it is imperative to understand the various sustainable manufacturing methods that can be incorporated to ensure the growth of circular fashion. The changes are made by various companies in their manufacturing methods and there are various methods carried out to utilise the manufactured products in later stages like reusing, recycling, and upcycling. In the case of manufacturing new products, companies implement Lean Six Sigma Quick Response (QR) (Just-in-Time [JIT]) and retail stores implement accurate response (AR) to ensure products are manufactured with minimal wastage as that is the first step to achieving circularity. It is because, if we receive the required amount of raw materials on time and not early, there is a reduction in inventory cost that includes holding cost and carrying cost thereby leading to reduced cost of the product. So just having materials delivered on time has such promising effects. Similarly, if the materials are used in an optimised manner, then the material cost which is approximately 60% of the total product cost can be reduced leading to the reduced cost of the product again. Along with this, if Lean Six Sigma and Right First Time (RFT) concepts are implemented, then the defects can be avoided, thereby reducing the wastage during pre-production and production of materials. Such measures will also avoid customer returns, thereby reducing post-production wastage. These measures in manufacturing lead to better utilisation of resources, reduced cost, and less load on the environment. and these factors contribute in turn to promoting circularity in fashion. Apart from this, in the case of manufacturing, there are methods like bio-fabrication and 3D printing, etc. which revolutionise the process reducing the wastage. In this section, such methods are discussed.

5.1 Bio-fabrication: Cultivating Sustainable Textiles

Bio-fabrication represents a paradigm shift in textile production. This approach utilises microorganisms or biological processes to create new materials. Here are some promising examples:

- Microbial Cellulose: This innovative material is derived from bacteria that are fed with sugar or other organic materials. The bacteria then produce cellulose, a natural polymer that can be processed into fibres for textile use. Microbial cellulose offers unique properties like enhanced strength, wrinkle resistance, and even biodegradability.
- Algae-Based Fabrics: Algae are fast-growing organisms that can be cultivated sustainably. These versatile organisms can be used to create bio-based fibres with a lower environmental footprint compared to traditional materials. Additionally,

algae have the added benefit of absorbing CO_2 during their growth, contributing to a cleaner environment.

- The Future of Bio-Fabrication: Bio-fabrication is still in its early stages, but the potential is vast. Researchers are exploring the use of various microorganisms and biological processes to create a diverse range of sustainable textiles. Challenges include scaling up production for commercial viability and ensuring the cost-effectiveness of these bio-based materials.

5.2 3D Printing and Digital Technologies: Transforming Production

Digital technologies are transforming the way fabrics are produced, offering significant sustainability benefits:

- 3D Printing for Fashion: 3D printing allows for on-demand production of garments, minimising waste typically generated during traditional cutting and sewing processes. This technology is particularly well suited for customised clothing, further reducing the need for mass production. Additionally, the creation of complex structures and designs is allowed in 3D printing, which is not possible in traditional methods.
- Digital Printing: Digital printing technologies use significantly less water compared to traditional printing methods that rely on large volumes of water and harsh chemicals. Digital printing also offers greater precision and allows for on-demand printing, reducing waste associated with excess fabric and overproduction.

5.3 Manufacturing Strategies

In this section, the various management strategies that aid in manufacturing with reduced waste are touched upon. The first one to be discussed is the quick response strategy. In this strategy, the materials are delivered to the manufacturer exactly before the start of the manufacturing by the supplier. It is also like sending the material at the right time to the retailer to start selling. If the material is sent earlier, then there will be inventory costs for the manufacturer or the retailer. The inventory was once considered to be an asset and now it is a liability as it involves various costs to hold more materials than required, including a loss in interest charges for the money invested in the holding of the raw materials. This cost is added to the final product, thereby increasing the cost of the product. Such costs can be reduced by using a QR strategy where the materials are sent when required which is called "Just in time" (JIT). Here the supplier retains the finished goods to supply at the right time and also the supplier acquires or produces the goods at the right time, thereby ensuring a proper supply chain process reducing wastage and promoting circularity in apparel and fashion. This concept also involves vendor-managed inventory (VMI) where there is a close-knit relationship between the stakeholders of the supply chain in which the vendor manages the inventory for the manufacturer till the required time and then sends it to the manufacturer or retailer for usage. These methods reduce the wastage at all places as the vendor works based on the order received by the manufacturer. A similar concept to this is the accurate response (AR) used in the retail industry where the products are manufactured based on demand. The normal staple items are manufactured beforehand and the fashion products are manufactured as per the demand. Since the fashion products demand is variable, enough allowance is kept to get those products ready and supplied quickly based on the requirement.

The other manufacturing concepts aiding circular fashion will be Lean Six Sigma and RFT concepts. The RFT means ensuring that the job is done correctly the first time, thereby avoiding wastage and improper utilisation of resources. It is difficult to implement RFT initially as there will be errors in manufacturing. However, proper training and increased awareness among workers will lead to its successful implementation. Apart from the RFT concept, Lean principles and the Six Sigma approach would also aid in circular fashion. The Six Sigma concept suggests that there can be 3.14 defects per million pieces produced and if an apparel company follows this principle, then it leads to reduced wastage and load on the environment. Also, Lean principles advocate the process of reducing seven types of waste in production, leading to better utilisation of resources and avoiding wastage of material and wastage of labour time. Such manufacturing strategies can surely aid in improving the circularity of fashion.

6 Designing for Longevity and Durability

6.1 Strategies for Creating Durable and Long-Lasting Apparel

In a world driven by fast fashion, where trends shift like dunes and clothes are seen as disposable items, the concept of durable apparel feels almost radical. But there is a growing movement towards quality over quantity and with good reason. Long-lasting clothes are not just better for your wallet but also for the environment (Cline, 2019). The longevity and durability of apparel depend on many factors such as fibre selection, yarn construction, fabric construction, construction and seam quality, care and maintenance. This chapter dives into the key strategies for creating apparel built to stand the test of time.

6.1.1 Fabric Fundamentals: Choosing for Longevity

- The foundation of any durable garment lies in its fabric. Natural fibres like organic cotton, linen, and wool are generally more durable than their synthetic counterparts. These materials breathe better, resist pilling (the formation of tiny balls of fibre on the fabric surface), and can often be repaired more easily.

- Cotton: The "go-to" natural fibre, cotton offers breathability, comfort, and washability, especially when opting for organic varieties (Cline 2019). Look for higher thread counts (indicating the number of threads per square inch) for a denser, more durable weave.
- Linen: Known for its strength and crisp drape, linen is a sustainable choice as it requires minimal water to grow. While it wrinkles easily, these wrinkles tend to fall out naturally with wear.
- Wool: A naturally insulating and moisture-wicking fibre, wool offers exceptional durability. Proper care, however (think gentle washing and air drying), is essential to maintain its quality.

6.1.2 Beyond Fibres: Construction Techniques for Strength

In apparel, stitches and seams play a major role, where stitches allow us to join the apparel components and seams allow us to give shape to the apparel. Stitches and seams with their performance properties add a great quality to the apparel. The strength and quality of the apparel depend on the physical and performance features (Akter & Khan, 2015). While fabric selection is crucial, construction methods also significantly impact garment longevity. Here are some key aspects to consider:

- Seams: Opt for flat-felled seams, which enclose raw fabric edges within the seam allowance, creating a stronger and smoother finish compared to basic seams.
  - French Seam: Mostly used for aesthetic purposes, this seam can be very strong when constructed correctly. It encloses all edges providing a clean and durable finish.
  - Welt Seam: Often used in outerwear, welt seams are durable and provide a clean, finished appearance.
  - Lapped Seam: While not as strong as the previous options, a lapped seam can be made durable by using appropriate fabric and reinforcing the seam with additional stitching (Gurarda, 2019).
- Stitching: Look for garments with dense stitching (more stitches per inch) for added strength, particularly at stress points like pockets and hems.
  - Long stitch provides a strong and durable seam.
  - Chain stitch is not as strong as lock stitch, but it is still quite durable and is often used in jeans and other heavy-duty garments.
  - Overlock stitch is used for finishing seams; it can also be used for construction. A four-thread overlock stitch provides good strength.
  - Safety stitch is a variation of the chain stitch that offers increased strength due to its double-needle construction.
  - Triple stitch involves sewing the same seam three times, providing exceptional strength.
- Reinforcement: Reinforcements on areas prone to wear and tear, like elbows on jackets or knees on pants, significantly extend the garment's lifespan.

6.1.3 Design for Durability: Timeless Styles and Mindful Details

Classic silhouettes and clean lines tend to stay relevant longer than trend-driven pieces. One can consider investing in well-tailored staples that can be easily combined and matched to create a variety of looks.

- Classic Cuts: A well-tailored blazer, a crisp white shirt, or a pair of dark wash jeans are enduring fashion styles that never go out of style and can be worn for years again and again.
- Multifunctional Garments: Opt for clothes that can be dressed up or down, maximising their versatility in your wardrobe.
- Minimal Hardware: Zippers, buttons, and other embellishments can be points of failure. Opt for minimalist designs or those with high-quality hardware.

6.1.4 Beyond the Purchase: Proper Care for Long Life

Investing in quality garments is just the first step. Proper care is essential to ensure they reach their full lifespan. Always follow the care instructions on the label and consider these additional tips:

- Wash Less, Wash Smarter: Overwashing can damage clothes. Spot clean whenever possible and wash only, when necessary, on a gentle cycle with cold water.
- Air Dry Whenever Possible: Heat is a major enemy of clothing. Skip the dryer whenever possible and opt for air drying to prevent shrinkage and fading.
- Proper Storage: Hang clothes on sturdy hangers to maintain their shape. Fold knitwear carefully to avoid stretching. Store delicate items in breathable storage bags.
- By following these strategies, you can build a wardrobe filled with durable, long-lasting pieces that you will love for years to come. Remember, quality over quantity is a sustainable approach to fashion; this benefits not just your wallet but also the planet.
- Stain Removal:
  - Act quickly, the sooner the stain is removed, the better.
  - Test the stain remover in an inconspicuous area first.
  - Use appropriate stain removers, as different stains require different treatments.
  - Avoid excessive rubbing as it can damage the fabric.

  Proper Care for Specific Fabric:

- Denim:
  - Wash inside out in cold water.
  - Avoid using fabric softener.
  - Air dry or tumble dry on low heat.
  - Avoid excessive washing.

- Synthetic Fabric.
  - Wash in warm or cold water.
  - Use a gentle detergent.
  - Avoid high heat when drying.
  - Examples: polyester, nylon, acrylic.
- Delicate Fabrics:
  - Hand wash or use a gentle cycle.
  - Avoid bleach and harsh detergents.
  - Dry flat or hang to dry.
  - Examples: silk, lace, cashmere, wool.

6.2 Role of Timeless Design in Sustainable Fashion

6.2.1 The Enduring Appeal: Timeless Design in Sustainable Fashion

In today's environmentally conscious fashion landscape, the concept of "timeless design" takes centre stage. It is a philosophy that transcends fleeting trends, prioritising quality, versatility, and a garment's ability to transcend seasons. This chapter explores the crucial role of timeless design in fostering a more sustainable fashion industry.

6.2.2 Reducing the Fast Fashion Trend

Fast fashion, characterised by its rapid production cycles and trend-driven designs, contributes significantly to environmental degradation and textile waste. Clothes are often made with cheap, synthetic materials, designed to be discarded quickly in favour of the next "it" item. Timeless design offers a powerful antidote to this unsustainable model.

- Reduced Production: By prioritising classic styles that remain relevant for years, the need for constant new collections diminishes. This results in lower production volumes, reducing the environmental impact that is associated with manufacturing processes (Cline, 2013).
- Lower Material Consumption: A shift towards timeless design encourages investment in fewer, higher-quality garments. This reduces the overall amount of fabric and resources needed to satisfy consumer demand.
- Combating Textile Waste: Timeless pieces are more likely to be cherished and worn for extended periods. This reduces textile waste which comes from discarded clothing that ends up in trash sites.

6.2.3 The Power of Versatility: Building a Wardrobe for Life

Timeless design goes beyond just classic silhouettes. It is about creating garments with versatility in mind, pieces that can be easily selected to form a part of any particular wear to create a range of different looks.

- Capsule Wardrobes: By focusing on core staples like tailored trousers, crisp shirts, and well-fitting jackets, one can build a capsule wardrobe that offers endless outfit combinations without succumbing to fleeting trends.
- Investment Pieces: Investing in high-quality, versatile pieces, like a timeless trench coat or a perfectly fitting blazer, creates a foundation for a long-lasting wardrobe.
- Neutral Colour Palettes: Opting for neutral colour palettes like black, navy, and beige allows for easier mixing and matching, extending the wearability of each garment.

6.2.4 Beyond Aesthetics: Quality Matters

Timeless design is not just about aesthetics; it is about prioritising quality construction for lasting wear. Here is why:

- Durable Fabrics: Natural fibres like organic cotton, linen, and wool are generally more durable than synthetics. They tend to drape better, resist pilling, and can often be repaired more easily.
- Superior Craftsmanship: Look for garments with strong seams, dense stitching, and reinforcements in high-stress areas. These elements contribute significantly to a garment's longevity.
- Classic Tailoring: Proper tailoring ensures a flattering fit and prevents garments from looking dated quickly. A well-tailored piece can be easily altered over time to accommodate slight changes in body shape.

6.2.5 The Emotional Connection: Loving Clothes That Stay in Your Heart

The timeless design fosters a deeper connection with our clothing. When we invest in quality pieces that we love, we are more likely to take care of them and wear them for years to come.

- Slow Fashion Philosophy: Timeless design aligns with the slow fashion philosophy, which emphasises quality over quantity and encourages a more mindful approach to consumption.
- Personal Style Evolution: Classic pieces can be a canvas for expressing personal style. By accessorising and layering timeless garments, you can create unique looks that reflect your individuality.

- Sustainable Investment: Investing in high-quality, timeless pieces can be seen as an investment in your style and the environment. These garments become cherished companions, not disposable trends.

In conclusion, timeless design serves as a cornerstone of sustainable fashion. By prioritising quality, versatility, and classic styles, we can move away from the wasteful practices of fast fashion and cultivate a more mindful relationship with our clothing. Timeless design allows us to build wardrobes that are not only stylish but also environmentally responsible, ensuring our clothes continue to tell our stories for years to come.

7 Conclusion

The fashion industry is at a very important phase. Its immense environmental impact demands urgent, transformative change. Circular fashion offers a promising path forward by prioritising waste reduction, efficient resource use, and extending the life of garments. Understanding the historical evolution of fashion's environmental challenges is essential. From the rise of synthetic fibres to the fast fashion phenomenon, the industry has left a significant ecological footprint. The shift towards sustainable materials and innovative production methods signals a growing awareness of the need for change. Circular fashion is the concept of designing for longevity. By focusing on durable fabrics, robust construction, and timeless styles, we can significantly reduce the industry's reliance on constant new production. However, achieving a truly circular fashion system requires a collective effort. Consumers, designers, manufacturers, and policymakers must all play their part. While challenges remain, the potential benefits of a circular fashion economy are immense. By working together, we can create a fashion industry that not only thrives economically but also respects the planet. We must prioritise policy changes, invest in innovative technologies, educate consumers, and foster strong collaborations. Only through a combined effort can we build a fashion industry that is both stylish and sustainable, and a small attempt is made in this chapter to detail the materials, methods, and designs that can be incorporated to contribute to sustainability using circular fashion concepts.

References

Akter, M., & Khan, M. R. (2015). The effect of stitch types and sewing thread types on seam strength for cotton apparel. *International Journal of Scientific & Engineering Research, 6*(7), 198–205.

Bell, J. (2023). *31 sustainable fabrics for the most eco-friendly fashion* [Online] Available from https://www.sustainablejungle.com/sustainable-fashion/sustainable-fabrics. Accessed on 24 June 2024.

Bismar, A. (2020). *Sixteen principles for a circular fashion industry* [Online]. Available from https://greenstrategy.se/produkt/circular-fashion-background-definition-and-16-key-principles/. Accessed on 20 June 2024.

Carter, C. R., & Rogers, D. S. (2008). A framework of sustainable supply chain management: Moving toward new theory. *International Journal of Physical Distribution & Logistics Management, 38*(5), 360–387.

Cline, E. L. (2013). *Overdressed: The shockingly high cost of cheap fashion*. Penguin.

Cline, E. L. (2019). *The conscious closet: The revolutionary guide to looking good while doing good*. Penguin.

Fransen, B. (2023). *The circular fashion economy: A sustainable shift towards a greener future* [Online]. Available from https://www.ecomatcher.com/the-circular-fashion-economy-a-sustainable-shift-towards-a-greener-future/. Accessed on 20 June 2024.

Gina-Marie, C. (2016). *The high environmental cost of fast fashion* [Online]. Available from https://www.triplepundit.com/2016/12/high-environmental-cost-fast-fashion/. Accessed on 20 June 2024.

Gurarda, A. (2019). Seam performance of garments. In *Textile manufacturing processes*. IntechOpen.

Kirchherr, J., Reike, D., & Hekkert, M. (2017). Conceptualizing the circular economy: An analysis of 114 definitions. *Resources, Conservation and Recycling, 127*, 221–232.

Nayak, R. (Ed.). (2022). *Sustainable fibres for fashion and textile manufacturing*. Woodhead Publishing.

Priyadarshini, M. B. I., & Prasad, U. D. (2024). From fast fashion to sustainable innovation: Managing the social and environmental impact of fashion tech. *International Journal of Engineering Technology and Management Sciences, 8*(2). https://doi.org/10.46647/ijetms.2024.v08i02.005

Raman, J., Kim, D. S., Kim, H. S., Oh, D. S., & Shin, H. J. (2022). Mycofabrication of mycelium-based leather from brown-rot fungi. *Journal of Fungi, 8*(3), 317.

Rathinamoorthy, R. (2019). Circular fashion. In *Circular economy in textiles and apparel* (pp. 13–48). Woodhead Publishing.

Index

S. S. Muthu (ed.), *Circular Business Models in the Apparel Industry*, Environmental Footprints and Eco-design of Products and Processes, https://doi.org/10.1007/978-3-031-80193-8

The manufacturer's authorised representative in the EU is Springer Nature Customer Service Centre GmbH, Europaplatz 3, 69115 Heidelberg, Germany. If you have any concerns regarding our products, please contact ProductSafety@springernature.com

Printed and bound by CPI Group (UK) Ltd, Croydon, CR0 4YY
01/10/2026
02225945-0009